Quantum Mechanics Simulations
The Consortium for Upper-Level Physics Software

John R. Hiller
Department of Physics, University of Minnesota
Duluth, Minnesota

Ian D. Johnston
Department of Physics, University of Sydney
Sydney, Australia

Daniel F. Styer
Department of Physics, Oberlin College
Oberlin, Ohio

Series Editors

William MacDonald

Maria Dworzecka

Robert Ehrlich

JOHN WILEY & SONS, INC.

NEW YORK · CHICHESTER · BRISBANE · TORONTO · SINGAPORE

ACQUISITIONS EDITOR Cliff Mills
MARKETING MANAGER Catherine Faduska
SENIOR PRODUCTION EDITOR Sandra Russell
DESIGNER Maddy Lesure
MANUFACTURING MANAGER Susan Stetzer

This book was set in 10/12 Times Roman by Beacon Graphics and
printed and bound by Hamilton Printing. The cover was printed by Phoenix Color.

Library of Congress Cataloging in Publication Data:
Hiller, John R.
    Quantum mechanics simulations : the consortium for upper level
physics software / John R. Hiller, Ian D. Johnson, Daniel F. Styer.
        p.   cm.
    Includes bibliographical references.
    ISBN 0-471-54884-7 (paper/disk)
    1. Physics—Data processing.   2. Physics—Computer programs.
  3. Consortium for Upper Level Physics Software.   I. Johnson, Ian D.
II. Styer, Daniel F.   III. Title.
QC52.H55   1994
530.1′2′01135365—dc20   94-38847
                                                               CIP

Printed in the United States of America

10   9   8   7   6   5   4   3   2

Contents

List of Figures

Quantum Mechanics Simulations
The Consortium for Upper-Level Physics Software

1

Introduction

"It is nice to know that the computer understands the problem. But I would like to understand it too."

—Eugene P. Wigner, quoted in *Physics Today,* July 1993

1.1 Using the Book and Software

The simulations in this book aim to exploit the capabilities of personal computers and provide instructors and students with valuable new opportunities to teach and learn physics, and help develop that all-important, if somewhat elusive, physical intuition. This book and the accompanying diskettes are intended to be used as supplementary materials for a junior- or senior-level course. Although you may find that you can run the programs without reading the text, the book is helpful for understanding the underlying physics, and provides numerous suggestions on ways to use the programs. *If you want a quick guided tour through the programs, consult the "Walk Throughs" in Appendix A.* The individual chapters and computer programs cover mainstream topics found in most textbooks. However, because the book is intended to be a supplementary text, no attempt has been made to cover all the topics one might encounter in a primary text.

Because of the book's organization, students or instructors may wish to deal with different chapters as they come up in the course, rather than reading the chapters in the order presented. One price of making the chapters semi-independent of one another is that they may not be entirely consistent in notation or tightly cross-referenced. Use of the book may vary according to the taste of the student or instructor. Students may use this material as the basis of a self-study course. Some instructors may make homework assignments from the large number of exercises in each chapter or to use them as the basis of student projects. Other instructors may use the computer programs primarily for in-class demonstrations. In this latter case, you may find that the programs are suitable for a range of courses from the introductory to the graduate level.

Use of the book and software may also vary with the degree of computer programming performed by users. For those without programming experience, all the computer simulations have been supplied in executable form, permitting them to be used as is. On the other hand, Pascal source code for the programs has also been provided, and a number of exercises suggest specific ways the programs can be modified. Possible modifications range from altering a single procedure especially set up for this purpose by the author, to larger modifications following given examples, to extensive additions for ambitious projects. However, the intent of the authors is that the simulations will help the student to develop intuition and a deeper understanding of the physics, rather than to develop computational skills.

We use the term "simulations" to refer to the computer programs described in the book. This term is meant to imply that programs include complex, often realistic, calculations of models of various physical systems, and the output is usually presented in the form of graphical (often animated) displays. Many of the simulations can produce numerical output—sometimes in the form of output files that could be analyzed by other programs. The user generally may vary many parameters of the system, and interact with it in other ways, so as to study its behavior in real time. The use of the term simulation should not convey the idea that the programs are bypassing the necessary physics calculations and simply producing images that look more or less like the real thing.

The programs accompanying this book can be used in a way that complements, rather than displaces, the analytical work in the course. It is our belief that, in general, computational and analytical approaches to physics can be mutually reinforcing. It may require considerable analytical work, for example, to modify the programs, or really to understand the results of a simulation. In fact, one important use of the simulations is to suggest conjectures that may then be verified, modified, or proven false analytically. A complete list of programs is given in Section 1.7.

1.2   Required Hardware and Installation of Programs

The programs described in this book have been written in the Pascal language for MS-DOS platforms. The language is Borland/Turbo Pascal, and the minimum hardware configuration is an IBM-compatible 386-level machine preferably with math coprocessor, mouse, and VGA color monitor. In order to accommodate a wide range of machine speeds, most programs that use animation include the capability to slow down or speed up the program. To install the programs, place disk number 1 in a floppy drive. Change to that drive, and type Install. You need only type in the file name to execute the program. Alternatively, you could type the name of the driver program (the same name as the directory in which the programs reside), and select programs from a menu. A number of programs write to temporary files, so you should check to see if your autoexec.bat file has a line that sets a temporary directory, such as SET TEMP = C:\TEMP. (If you have installed WINDOWS on your PC, you will find that such a command has already been written into your autoexec.bat file.) If no such line is there, you should add one.

Compilation of Programs

If you need to compile the programs, it would be preferable to do so using the Borland 7.0 (or later) compiler. If you use an earlier Turbo compiler you may run out of memory when compiling. If that happens, try compiling after turning off memory resident programs. If your machine has one, be sure to compile with the math-coprocessor turned on (no emulation). Finally, if you recompile programs using any compiler other than Borland 7.0, you will get the message: "EGA/VGA Invalid Driver File" when you try to execute them, because the driver file supplied was produced using this version of the compiler. In this case, search for the file BGILINK.pas included as part of the compiler to find information on how to create the EGAVGA.obj driver file. *If any other instructions are needed for installation, compilation, or running of the programs, they will be given in a README file on the diskettes.*

1.3   User Interface

To start a program, simply type the name of the individual or driver program, and an opening screen will appear. All the programs in this book have a common user interface. Both keyboard and mouse interactions with the computer are possible. Here are some conventions common to all the programs.

Menus: If using the *keyboard*, press **F10** to highlight one of menu boxes, then use the **arrow** keys, **Home**, and **End** to move around. When you press **Return** a submenu will pull down from the currently highlighted menu option. Use the same keys to move around in the submenu, and press **Return** to choose the highlighted submenu entry. Press **Esc** if you want to leave the menu without making any choices.

If using the *mouse* to access the top menu, click on the menu bar to pull down a submenu, and then on the option you want to choose. Click anywhere outside the menus if you want to leave them without making any choice. Throughout this book, the process of choosing submenu entry **Sub** under main menu entry **Main** is referred to by the phrase "choose **Main | Sub**." The detailed structure of the menu will vary from program to program, but all will contain **File** as the first (left-most) entry, and under **File** you will find **About CUPS, About Program, Configuration,** and **Exit Program**. The first two items when activated by mouse or arrows keys will produce information screens. Selecting **Exit Program** will cause the program to terminate, and choosing **Configuration** will present you with a list of choices (described later), concerning the mode of running the program. In addition to these four items under the **File** menu, some programs may have additional items, such as **Open**, used to open a file for input, and **Save**, used to save an output file. If **Open** is present and is chosen, you will be presented with a scrollable list of files in the current directory from which to choose.

Hot Keys: Hot keys, usually listed on a bar at the bottom of the screen, can be activated by pressing the indicated key or by clicking on the hot key bar with the mouse. The hot key **F1** is reserved for help, the hot key **F10** activates the menu bar. Other hot keys may be available.

Sliders (scroll-bars): If using the *keyboard*, press **arrow** keys for slow scrolling of the slider, **PgUp/PgDn** for fast scrolling, and **End/Home** for moving from one end to another. If you have more then one slider on the screen then only the slider with marked "thumb" (sliding part) will respond to the above keys. You can toggle the mark between your sliders by pressing the **Tab** key.

If using the *mouse* to adjust a slider, click on the thumb of the slider, drag it to desired value, and release. Click on the arrow on either end of the slider for slow scrolling, or in the area on either side of thumb for fast scrolling in this direction. Also, you can click on the box where the value of the slider is displayed, and simply type in the desired number.

Input Screens: All input screens have a set of "default" values entered for all parameters, so that you can, if you wish, run the program by using these original values. Input screens may include circular radio buttons and square check boxes, both of which can take on Boolean, i.e., "on" or "off," values. Normally, check boxes are used when only one can be chosen, and radio buttons when any number can be chosen.

If using the *keyboard*, press **Return** to accept the screen, or **Esc** to cancel it and lose the changes you may have made. To make changes on the input screen by keyboard, use **arrow** keys, **PgUp**, **PgDn**, **End**, **Home**, **Tab**, and **Shift-Tab** to choose the field you want to change, and use the backspace or delete keys to delete numbers. For Boolean fields, i.e., those that may assume one of two values, use any key except those listed above to change its value to the opposite value.

If you use the *mouse*, click [OK] to accept the screen or [Cancel] to cancel the screen and lose the changes. Use the mouse to choose the field you want to change. Clicking on the Boolean field automatically changes its value to the opposite value.

Parser: Many programs allow the user to enter expressions of one or more variables that are evaluated by the program. The function parser can recognize the following functions: absolute value (abs), exponential (exp), integer or fractional part of a real number (int or frac), real or imaginary part of a complex number (re or im), square or square root of a number (sqr or sqrt), logarithms—base 10 or e (log or ln)—unit step function (h), and the sign of a real number (sgn). It can also recognize the following trigonometric functions: sin, cos, tan, cot, sec, csc, and the inverse functions arcsin, arccos, arctan, as well as the hyperbolic functions denoted by adding an "h" at the end of all the preceding functions. In addition, the parser can recognize the constants pi, e, $i(\sqrt{-1})$, and rand (a random number between 0 and 1). The operations **+**, **−**, *****, **/**, ^(exponentiation), and !(factorial) can all be used, and the variables r and c are interpreted as $r = \sqrt{x^2 + y^2}$ and $c = x + iy$. Expressions involving these functions, variables, and constants can be nested to an arbitrary level using parentheses and brackets. For example, suppose you entered the following expression: **h(abs(sin(10*pi* x))−0.5)**. The parser would interpret this function as $h(|sin(10\pi x)|-0.5)$. If the program evaluates this function for a range of x-values, the result, in this case, would be a series of square pulses of width 1/15, and center-to-center separation 1/10.

Help: Most programs have context-sensitive help available by pressing the **F1** hot
key (or clicking the mouse in the **F1** hot key bar). In some programs help is
also available by choosing appropriate items on the menu, and in still other
programs tutorials on various aspects of the program are available.

1.4   The CUPS Project and CUPS Utilities

The authors of this book have developed their programs and text as part of the
Consortium for Upper-Level Physics Software (CUPS). Under the direction of the
three editors of this book, CUPS is developing computer simulations and associated texts for nine junior- or senior-level courses, which comprise most of the undergraduate physics major curriculum during those two years. A list of the nine
CUPS courses, and the authors associated with each course, follows this section.
This international group of 27 physicists includes individuals with extensive back-
grounds in research, teaching, and development of instructional software.

The fact that each chapter of the book has been written by a different author
means that the chapters will reflect that individual's style and philosophy. Every
attempt has been made by the editors to enhance the similarity of chapters, and
to provide a similar user interface in each of the associated computer simulations. Consequently, you will find that the programs described in this and other
CUPS books have a common look and feel. This degree of similarity was made
possible by producing the software in a large group that shared a common philosophy and commitment to excellence.

Another crucial factor in developing a degree of similarity between all CUPS
programs is the use of a common set of utilities. These CUPS utilities were written
by Jaroslaw Tuszynski and William MacDonald, the former having responsibility
for the graphics units, and the latter for the numerical procedures and functions.
The numerical algorithms are of high quality and precision, as required for reliable results. CUPS utilities were originally based on the M.U.P.P.E.T. utilities of
Jack Wilson and E.F. Redish, which provided a framework for a much expanded
and enhanced mathematical and graphics library. The CUPS utilities (whose
source code is included with the simulations with this book), include additional
object-oriented programs for a complete graphical user interface, including pull-
down menus, sliders, buttons, hotkeys, and mouse clicking and dragging. They
also include routines for creating contour, two-dimensional (2-D) and 3-D plots,
and a function parser. The CUPS utilities have been provided in source code form
to enable users to run the simulations under future generations of Borland/Turbo
Pascal. If you do run under future generations of Turbo or Borland Pascal on the
PC, the utilities and programs will need to be recompiled. You will also need
to create a new egavga.obj file which gets combined with the programs when an
executable version is created—thereby avoiding the need to have separate
(egavga.bgi) driver files. These CUPS utilities are also available to users who
wish to use them for their own projects.

One element not included in the utilities is a procedure for creating hard copy
based on screen images. When hard copy is desired, those PC users with the appropriate graphics driver (graphics.com), may be able to produce high-quality
screen images by depressing the **PrintScreen** key. If you do not have the graph-
ics software installed to get screen dumps, select **Configuration | Print Screen**,

and follow the directions. Moreover, public domain software also exists for capturing screen images, and for producing PostScript files, but the user should be aware that such files are often quite large, sometimes over 1 MB, and they require a PostScript printer driver to produce.

One feature of the CUPS utilities that can improve the quality of hard copy produced from screen captures is a procedure for switching colors. This capability is important because the gray scale rendering of colors on black-and-white printers may create poor contrasts if the original (default) color assignments are used. To access the CUPS utility for changing colors, the user need only choose **Configuration** under the **File** menu when the program is first initiated, or at any later time. Once you have chosen **Configuration**, to change colors you need to click the mouse on the **Change Colors** bar, and you will be presented with a 16 by 16 matrix of radio buttons that will allow you to change any color to any other color, or else to use predefined color switches, such as a color "reversal," or a conversion of all light colors to black, and all dark colors to white. (The screen captures given in this book were produced using the "reverse" color map.) Any such color changes must be redone when the program is restarted.

Other system parameters may likewise be set from the **File | Configuration** menu item. These include the path for temporary files that the program may create (or want to read), the mouse "double click" speed—important for those with slow reflexes—an added time delay to slow down programs on computers that are too fast, and a "check memory" option—primarily of interest to those making program modifications.

Those users wishing more information on the CUPS utilities should consult the CUPS Utilities Manual, written by Jaroslaw Tuszynski and William MacDonald, published by John Wiley and Sons. However, it is not necessary for casual users of CUPS programs to become familiar with the utilities. Such familiarity would only be important to someone wishing to write their own simulations using the utilities. The utilities are freely available for this purpose, for unrestricted noncommercial production and distribution of programs. However, users of the utilities who wish to write programs for commercial distribution should contact John Wiley and Sons.

1.5   Communicating With the Authors

Users of these programs should not expect that run-time errors will never occur! In most cases, such run-time errors may require only that the user restart the program; but in other cases, it may be necessary to reboot the computer, or even turn it off and on. The causes of such run-time errors are highly varied. In some cases, the program may be telling you something important about the physics or the numerical method. For example, you may be trying to use a numerical method beyond its range of applicability. Other types of run-time errors may have to do with memory or other limitations of your computer. Finally, although the programs in this book have been extensively tested, we cannot rule out the possibility that they may contain errors. (Please let us know if you find any! It would be most helpful if such problems were communicated by electronic mail, and with complete specificity as to the circumstances under which they arise.)

It would be best if you communicated such problems directly to the author of each program, and simultaneously to the editors of this book (the CUPS Directors), via electronic mail—see addresses listed below. Please feel free to communicate any suggestions about the programs and text which may lead to improvements in future editions. Since the programs have been provided in source code form, it will be possible for you to make corrections of any errors that you or we find in the future—provided that you send in the registration card at the back of the book, so that you can be notified. The fact that you have the source code will also allow you to make modifications and extensions of the programs. We can assume no responsibility for errors that arise in programs that you have modified. In fact, we strongly urge you to change the program name, and to add a documentary note at the beginning of the code of any modified programs that alerts other potential users of any such changes.

1.6   CUPS Courses and Developers

- **CUPS Directors**
  Maria Dworzecka, George Mason University (cups@gmuvax.gmu.edu)
  Robert Ehrlich, George Mason University (cups@gmuvax.gmu.edu)
  William MacDonald, University of Maryland (w_macdonald@umail.umd.edu)

- **Astrophysics**
  J. M. Anthony Danby, North Carolina State University (n38hs901@ncuvm.ncsu.edu)
  Richard Kouzes, Battelle Pacific Northwest Laboratory (rt_kouzes@pnl.gov)
  Charles Whitney, Harvard University (whitney@cfa.harvard.edu)

- **Classical Mechanics**
  Bruce Hawkins, Smith College (bhawkins@smith.bitnet)
  Randall Jones, Loyola College (rsj@loyvax.bitnet)

- **Electricity and Magnetism**
  Robert Ehrlich, George Mason University (rehrlich@gmuvax.gmu.edu)
  Lyle Roelofs, Haverford College (lroelofs@haverford.edu)
  Ronald Stoner, Bowling Green University (stoner@andy.bgsu.edu)
  Jaroslaw Tuszynski, George Mason University (cups@gmuvax.gmu.edu)

- **Modern Physics**
  Douglas Brandt, Eastern Illinois University (cfdeb@ux1.cts.eiu.edu)
  John Hiller, University of Minnesota, Duluth (jhiller@d.umn.edu)
  Michael Moloney, Rose Hulman Institute (moloney@nextwork.rose-hulman.edu)

- **Nuclear and Particle Physics**
  Roberta Bigelow, Willamette University (rbigelow@willamette.edu)
  John Philpott, Florida State University (philpott@fsunuc.physics.fsu.edu)
  Joseph Rothberg, University of Washington (rothberg@phast.phys.washington.edu)

- **Quantum Mechanics**
  John Hiller, University of Minnesota, Duluth (jhiller@d.umn.edu)
  Ian Johnston, University of Sydney (idj@suphys.physics.su.oz.au)
  Daniel Styer, Oberlin College (dstyer@physics.oberlin.edu)

- **Solid State Physics**
  Graham Keeler, University of Salford (g.j.keeler@sysb.salford.ac.uk)
  Roger Rollins, Ohio University (rollins@chaos.phy.ohiou.edu)
  Steven Spicklemire, University of Indianapolis (steves@truevision.com)

- **Thermal and Statistical Physics**
  Harvey Gould, Clark University (hgould@vax.clarku.edu)
  Lynna Spornick, Johns Hopkins University
  Jan Tobochnik, Kalamazoo College (jant@kzoo.edu)

- **Waves and Optics**
  G. Andrew Antonelli, Wolfgang Christian, and Susan Fischer, Davidson College (wc@phyhost.davidson.edu)
  Robin Giles, Brandon University (giles@brandonu.ca)
  Brian James, Salford University (b.w.james@sysb.salford.ac.uk)

1.7   Descriptions of all CUPS Programs

Each of the computer simulations in this book (as well as those in the eight other books comprised by the CUPS Project) are described below. The individual headings under which programs appear correspond to the nine CUPS courses. In several cases, programs are listed under more than one course. The number of programs listed under the Astrophysics, Modern Physics, and Thermal Physics courses is appreciably greater than the others, because several authors have opted to subdivide their programs into many smaller programs. Detailed inquiries regarding CUPS programs should be sent to the program authors.

ASTROPHYSICS PROGRAMS

STELLAR (Stellar Models), written by Richard Kouzes, is a simulation of the structure of a static star in hydrodynamic equilibrium. This provides a model of a zero age main sequence star, and helps the user understand the physical processes that exist in stars, including how density, temperature, and luminosity depend on mass. Stars are self-gravitating masses of hot gas supported by thermodynamic processes fueled by nuclear fusion at their core. The model integrates the four differential equations governing the physics of the star to reach an equilibrium condition which depends only on the star's mass and composition.

EVOLVE (Stellar Evolution), written by Richard Kouzes, builds on the physics of a static star, and considers (1) how a gas cloud collapses to become a main sequence star, and (2) how a star evolves from the main sequence to its final demise. The model is based on the same physics as the STELLAR program. Starting from a diffuse cloud of gas, a protostar forms as the cloud collapses and reaches a sufficient density for fusion to begin. Once a star reaches equilibrium, it remains for most of its life on the main sequence, evolving off after it has consumed its fuel. The final stages of the star's life are marked by rapid and dramatic evolution.

BINARIES is the driver program for all Binaries programs (**VISUAL1, VISUAL2, ECLIPSE, SPECTRO, TIDAL, ROCHE, and ACCRDISK**).

VISUAL1 (Visual Binaries—Proper Motion), written by Anthony Danby, enables you to visualize the proper motion in the sky of the members of a visual binary system. You can enter the elements of the system and the mass ratio, as well as the speed at which the center of mass moves across the screen. The program also includes an animated three-dimensional demonstration of the elements.

VISUAL2 (Visual Binaries—True Orbit), written by Anthony Danby, enables you to select an apparent orbit for the secondary star with arbitrary eccentricity, with the primary at any interior point. The elements of the orbit are displayed. You can see the orbit animated in three dimensions, or can make up a set of "observations" based on the apparent orbit.

ECLIPSE (Eclipsing Binaries), written by Anthony Danby, shows simultaneously either the light curve and the orbital motion or the light curve and an animation of the eclipses. You can select the elements of the orbit and radii and magnitudes of the stars. A form of limb-darkening is also included as an option.

SPECTRO (Spectroscopic Binaries), written by Anthony Danby, allows you to select the orbital elements of a spectroscopic binary, and then shows simultaneously the velocity curve, the orbital motion, and a moving spectral line.

TIDAL (Tidal Distortion of a Binary), written by Anthony Danby, models the motion of a spherical secondary star around a primary that is tidally distorted by the secondary. You can select orbital elements, masses of the stars, a parameter describing the tidal lag, and the initial rate of rotation of the primary. The equations are integrated over a time interval that you specify. Then you can see the changes of the orbital elements, and the rotation of the primary, with time. You can follow the motion in detail around each revolution, or in a form where the equations have been averaged around each revolution.

ROCHE (The Photo-Gravitational Restricted Problem of Three Bodies), written by Anthony Danby, follows the two-dimensional motion of a particle that is subject to the gravitational attraction of two bodies in mutual circular orbits, and also, optionally, radiation pressure from these bodies. It is intended, in part, as background for the interpretation of the formation of accretion disks. Curves of zero velocity (that limit regions of possible motion) can be seen. The orbits can also be followed using Poincaré maps.

ACCRDISK (Formation of an Accretion Disk), written by Anthony Danby, follows some of the dynamical steps in this process. The dynamics is valid up to the initial formation of a hot spot, and qualititative afterward.

NBMENU is the driver program for all programs on the motion of N interacting bodies: **TWOGALAX, ASTROIDS, N-BODIES, PLANETS, PLAYBACK, and ELEMENTS**.

TWOGALAX (The Model of Wright and Toomres), written by Anthony Danby, is concerned with the interaction of two galaxies. Each consists of a central gravitationally attracting point, surrounded by rings of stars (which are attracted, but do not attract). Elements of the orbits of one galaxy relative to the other are selected, as is the initial distribution and population of the rings. The motion can be viewed as projected into the plane of the orbit of the galaxies, or simultaneously in that plane and perpendicular to it. The positions can be stored in a file for later viewing.

ASTROIDS (N-Body Application to the Asteroids), written by Anthony Danby, uses the same basic model, but a planet and a star take the place of the galaxies and the asteroids replace the stars. Emphasis is on asteroids all having the same period, with interest on periods having commensurability with the period of the planet. The orbital motion of the system can be followed. The positions can be stored in a file for later viewing. An asteroid can be selected, and the variation of its orbital elements can then be followed.

NBODIES (The Motion of N Attracting Bodies), written by Anthony Danby, allows you to choose the number of bodies (up to 20) and the total energy of the system. Initial conditions are chosen at random, consistent with this energy, and the resulting motion can be observed. During the motion various quantities, such as the kinetic energy, are displayed. The positions can be stored in a file for later viewing.

PLANETS (Make Your Own Solar System), written by Anthony Danby, is similar to the preceding program, but with the bodies interpreted as a star with planets. Initial conditions are specified through the choice of the initial elements of the planets. The positions can be stored in a file for later viewing.

PLAYBACK, written by Anthony Danby, enables a file stored by one of the preceding programs to be viewed.

ELEMENTS (Orbital Elements of a Planet), written by Anthony Danby, shows a three-dimensional animation that can be viewed from any angle.

GALAXIES is the driver program for Galactic Kinematics programs: **ROTATION, OORTCONS, and ARMS21CM**.

ROTATION (The Rotation Curve of a Galaxy), written by Anthony Danby, first prompts you to "design" a galaxy, consisting of a central mass and up to five spheroids (that can be visible or invisible). It then displays the galaxy and can show the animated rotation or the rotation curve.

OORTCONS (Galactic Kinematics and Oort's Constants), written by Anthony Danby, allows you to design your galaxy, choose the location of the "sun" and a local region around it, and the to observe the kinematics in this region. It also shows graphs of radial velocity and proper motion in comparison with the linear approximation, and computes the Oort constants.

ARMS21CM (The Spiral Structure of a Galaxy), written by Anthony Danby, allows you to design your galaxy, construct a set of spiral arms, and select the position of the "sun." Then, for different galactic longitudes, you can see observed profiles of 21 cm lines.

ATMOS (Stellar Atmospheres), written by Charles Whitney, permits the user to select a constellation, see it mapped on the computer screen, point to a star, and see it plotted on a brightness-color diagram. The user's task is to build a model atmosphere that imitates the photometric properties of observed stars. This is done by specifying numerical values for three basic stellar parameters: radius, mass, and luminosity. The program then builds the model and displays it on the brightness-color diagram, and it also plots the spectrum and the detailed thermodynamic structure of the atmosphere. With this program the user may investigate the relation between stellar parameters and the thermal properties of the gas in the atmosphere. Two atmospheres may be superposed on the graphs, for easier comparison.

PULSE (Stellar Pulsations), written by Charles Whitney, illustrates stellar pulsation by simulating the thermo-mechanical behavior of a "star" modeled by a self-gravitating gas divided by spherical elastic shells. The elastic shells resemble a set of coupled oscillators. The program solves for the modes of small-amplitude motion, and it uses Fourier synthesis to construct motions for arbitrary starting conditions. The screen displays the thermodynamic structure and surface properties, such as temperature, pressure, and velocity. Animation displays the nature of the pulsation. By showing the motions, temperatures, and energy flux, the program demonstrates the heat engine acting inside the pulsating star. The motions of the shells and the spatial Fourier decomposition into eigenmodes are displayed simultaneously, and this will help you visualize the meaning of the Fourier components.

CLASSICAL MECHANICS PROGRAMS

GENMOT (The Motion Generator), written by Randall Jones, allows you to solve numerically any differential equation of motion for a system with up to three degrees of freedom and display the time evolution of the system in a wide variety of formats. Any of the dynamical variables or any function of those variables may be displayed graphically and/or numerically and a wide range of animations may be constructed. Since the Motion Generator can be used to solve any second-order differential equation, it can also be used to study systems analyzed by Lagrangian methods. Real world coordinates may be constructed as functions of generalized coordinates so that simulations of the actual system can be constructed.

ROTATE (Rotation of 3-D Objects), written by Randall Jones, is designed to aid in the visualization of the dynamical variables of rotational motion. It will allow you to observe the 3-D motion of rotating objects in a controlled fashion, running the simulation faster, slower, or in reverse while displaying the corresponding evolution of the angular velocity, the angular momentum and the torque. It will display the motion from the fixed frame and from the body frame to help in understanding the translation between these two descriptions of the motion. By using the stereographic feature of the program you can create a genuine 3-D representation of the motion of the quantities.

COUPOSC (Coupled Oscillators), written by Randall Jones, is designed to investigate a wide range of harmonic systems. Given a set of objects and springs connected in one or two dimensions, the simulation can solve the problem by generating the normal mode frequencies and their corresponding motions. It can take any set of initial conditions and resolve them into their component normal mode motions or take any set of initial mode occupations and display the corresponding motions of the objects. It can also determine the motion of the system when it is acted on by external forces. In this case the total forces are no longer harmonic, so the solution is generated numerically. The harmonic analysis, however, still provides an important tool for investigating and understanding the subsequent motion.

ANHARM (Anharmonic Oscillators), written by Bruce Hawkins, simulates oscillations of various types: pendulum, simple harmonic oscillator, asymmetric, cubic, Vanderpol, and a mass in the center of a spring with fixed ends. Nonlinear behavior is emphasized. The user may choose to view one to four graphs of the motion simultaneously, along with the potential diagram and a picture of the moving object. Graphs that may be viewed are x vs. t, v vs. t, v vs. x, the Poincaré diagram, and the return map. Tools are provided to explore parameter space for regions of interest. Fourier analysis is available, resonance diagrams can be plotted, and the period can be plotted as a function of energy. Includes a tutorial demonstrating the usefulness of phase plots and Poincaré plots.

ORBITER (Gravitational Orbits), written by Bruce Hawkins, simulates the motion of up to five objects with mutually gravitational attraction, and any reasonable number of additional objects moving in the gravitation field of the first five. The motion may be viewed in up to six windows simultaneously: windows centered on a particular body, on the center of mass, stationary in the universe frame, or rotating with the line joining the two most massive bodies. A menu of available systems includes the solar system, the sun/earth/moon system; the sun, Jupiter, and its moons; the sun, earth, and Saturn, demonstrating retrograde motion; the sun, Jupiter, and a comet; and a pair of binary stars with a comet. Bodies may be added to any system, or a new system created using either numerical coordinates or the mouse. Bodies may be replicated to demonstrate the sensitivity of orbits to initial conditions.

COLISION (Collisions), written by Bruce Hawkins, simulates two-body collisions under any of a number of force laws: Coulomb with and without shielding and truncation, hard sphere, soft sphere (harmonic), Yukawa, and Woods-Saxon. Collision may be viewed in the laboratory and center of mass systems simultaneously, with or without momentum diagrams. Includes a tutorial on the usefulness of the center of mass system, one on the kinematics of relativistic collisions, and one on cross section. Plots cross section against scattering parameter, and compares collisions at different parameters.

ELECTRICITY AND MAGNETISM PROGRAMS

FIELDS (Analysis of Vector and Scalar Fields), written by Jarek Tuszynski, displays scalar and vector fields for any algebraic or trigonometric expression entered by the user. It also computes numerically the divergence, curl, and Laplacian for the vector fields, and the gradient and Laplacian for the scalar fields. Simultaneous displays of selected quantities are provided in user-selected planes, using vector, contour, or 3-D plots. The program also allows the user to define paths along which line integrals are computed.

GAUSS (Gauss' Law), written by Jarek Tuszynski, treats continuous charge distributions having spherical or cylindrical symmetry, and those that vary as a function of the x-coordinate only. The program allows the user to enter an arbitrary function to define either the electric field magnitude, the potential, or the charge density. It then computes the other two functions by numerical differentiation or integration, and displays all three functions. Finally, the program allows the user to enter a "comparison function," which is plotted on the same graph, so as to check whether his analytic solutions are correct.

POISSON (Poisson's Equation Solved on a Grid), written by Jarek Tuszynski, solves Poisson's equation iteratively on a 2-D grid using the method of simultaneous over-relaxation. The user can draw arbitrary systems consisting of line charges, and charged conducting cylinders, plates, and wires, all infinite in extent perpendicular to the grid. After iteratively solving Poisson's equation, the program displays the results for the potential, electric field, or the charge density (found from the Laplacian of the potential), in the form of contour, vector, or 3-D plots. In addition, many other program features are available, including the ability to specify surfaces, along which the potential varies according to some algebraic function specified by the user.

IMAG&MUL (Image Charges and Multipole Expansion), written by Lyle Roelofs and Nathaniel Johnson, allows users to explore two approaches to the solution of Laplace's equation—the image charge method and expansion in multipole moments. In the image charge mode (IC) the user is presented with a variety of configurations involving conducting planes and point charges and is asked to "solve" each by placing image charges in the appropriate locations. The program displays the electric field due to all point charges, real and image, and a solution can be regarded as successful with the field due to all charges is everywhere orthogonal to all conducting surfaces. Solutions can then be examined with a variety of included software "tools." The multipole expansion (ME) mode of the program also permits a "hands-on" exploration of standard electrostatic problems, in this case the "exterior" problem, i.e., the determination of the field outside a specified equipotential surface. The program presents the user with a variety of azimuthally symmetric equipotential surfaces. The user "solves" for the full potential by adding chosen amounts of the (first six) multipole moments. The screen shows the contours of the summed potential and the problem is "solved" when the innermost contour matches the given equipotential surface as closely as possible.

ATOMPOL (Atomic Polarization), written by Lyle Roelofs and Nathaniel Johnson, is an exploration of the phenomenon of atomic polarization. Up to 36 atoms of controllable polarizability are immersed in an external electric field. The program solves for and displays the field throughout the region in which the atoms are located. A closeup window shows the polarization of selected atoms and software "tools" allow for further analysis of the resulting electric fields. Use of this program improves the student's understanding of polarization, the interaction of polarized entities, and the atomic origin of macroscopic polarization, the latter via study of closely spaced clusters of polarizable atoms.

DIELECT (Dielectric Materials), written by Lyle Roelofs and Nathaniel Johnson, is a simulation of the behavior of linear dielectric materials using a cell-based approach. The user controls either the polarization or the susceptibility of each cell in a (25 × 25) grid (with assumed uniformity in the third direction). Full self-consistent solutions are obtained via an iterative relaxation method and the fields P, E, or D are displayed. The student can investigate the self-interactions of polarized materials and many geometrical effects. Use of this program aids the student in developing understanding of the subtle relations among and meaning of P, E, and D.

ACCELQ (Fields From an Accelerated Charge), written by Ronald Stoner, simulates the electromagnetic fields in the plane of motion generated by a point charge that is moving and accelerating in two dimensions. The user chooses from among seven predefined trajectories, and sets the values of maximum speed and viewing time. The electric field pattern is recomputed after each change of trajectory or parameter; thereafter, the user can investigate the electric field, magnetic field, retarded potentials, and Poynting-vector field by using the mouse as a field probe, by using gridded overlays, or by generating plots of the various fields along cuts through the viewing plane.

QANIMATE (Fields From an Accelerated Charge—Animated Version), written by Ronald Stoner, is an interactive animation of the changing electric field pattern generated by a point electric charge moving in two dimensions. Charge motion can be manipulated by the user from the keyboard. The display can include electric field lines, radiation wave fronts, and their points of intersection. The motion of the charge is controlled by the using **arrow** keys to accelerate and steer much like the accelerator and steering wheel of a car, except that acceleration must be changed in increments, and the **Space** bar can used to engage or disengage the steering. With steering engaged, the charge will move in a circle. Unless the acceleration is made zero, the speed will increase (or decrease) to the maximum (minimum) possible value. At constant speed and turning rate, the charge can be controlled by the **Space** bar alone.

EMWAVE (Electromagnetic Waves), written by Ronald Stoner, uses animation to illustrate the behavior of electric and magnetic fields in a polarized plane electromagnetic wave. The user can choose to observe the wave in free space, or to see the effect on the wave of incidence on a material interface, or to see the effects of optical elements that change its polarization. The user can change the polarization state of the incident wave by specifying its Stokes parameters. Standing electromagnetic waves can be simulated by combining the incident traveling wave with a reflected wave of the same amplitude. The user can do that by choosing appropriate values of the physical properties of the medium on which the incident wave impinges in one of the animations.

MAGSTAT (Magnetostatics), written by Ronald Stoner, computes and displays magnetic fields in and near magnetized materials. The materials are uniform and have 3-D shapes that are solids of revolution about a vertical axis. The shape of the material can be modified or chosen from a data input screen. The user has the option of generating the fields produced by a permanently and uniformly magnetized object, or of generating the fields of a magnetizable object placed in an otherwise uniform external field. Besides choosing the shape and aspect ratio of the object, the user can vary the magnetic permeability of the magnetizable material, and choose among three fields to display: magnetic induction (B), magnetic field strength (H), and magnetization (M). Each of these fields can be displayed or explored in several different ways. The algorithm for computing the fields uses a superposition of Chebyschev polynomial approximants to the H field due to "rings" of "magnetic charge."

MODERN PHYSICS PROGRAMS

NUCLEAR (Nuclear Energetics and Nuclear Counting), written by Michael Moloney, deals with basic nuclear properties related to mass, charge, and energy, for approximately 1900 nuclides. Graphs are available involving binding energy, mass, and Q values of a variety of nuclear reactions, including alpha and beta decays. Part 2 deals with simulating the statistics of counting with a Geiger-Muller tube. This part also simulates neutron activation, and the counting behavior as neutron flux is turned on and off. Finally, a decay chain from A to B to C is simulated, where half-lives may be changed, and populations are graphed as a function of time.

GERMER (Davisson-Germer and G. P. Thomson Experiments), written by Michael Moloney, simulates both the Davisson-Germer and G. P. Thomson experiments with electrons scattering from crystalline materials. Stress is laid on the behavior of electrons as waves; similarities are noted with scattering of x-rays. The exercises encourage students to understand why peaks and valleys in scattered electrons occur where they do.

QUANTUM (one-dimensional Quantum Mechanics), written by Douglas Brandt, is a program that has four sections. The first section allows users to investigate the uncertainty principle for specified wavefunctions in position or momentum space. The second section allows users to investigate the time evolution of wavepackets under various dispersion relations. The third section allows users to investigate solutions to Schrödinger's equation for asymptotically free solutions. The user can input a barrier and the program calculates reflection and transmission coefficients for a range of energies and show wavepacket time evolution for the barrier potential. The fourth section is similar to the third, except that it allows the user to investigate bound solutions to Schrödinger's equation. The program calculates the bound state Hamiltonian eigenvalues and spatial eigenfunctions.

RUTHERFD (Rutherford Scattering), written by Douglas Brandt, is a program for investigating classical scattering of particles. A scattering potential can be chosen from a list of predefined potentials or an arbitrary potential can be input by the user. The computer generates scattering events by randomly picking impact parameters from a distribution defined by beam parameters specified by the user. It displays the results of the scattering on a polar histogram and on a detailed histogram to help users gain insight into differential scattering cross section. A scintillation mode can be chosen for users that want more appreciation of the actual experiments of Geiger and Marsden. A "guess the scatterer" mode is available for trying to gain appreciation of how scattering experiments are used to infer properties of the scatterers.

SPECREL (Special Relativity), written by Douglas Brandt, is a program to investigate special relativity. The first section is to investigate change of coordinate systems through Minkowski diagrams. The user can define coordinates of objects in one reference frame and the computer calculates the coordinates in a user-selectable coordinate system and displays the objects in both reference frames. The second section allows users to view clocks that are in relative motion. A clock can be given an arbitrary trajectory through space-time and the readings of various clocks can be viewed as the clock follows that trajectory. A third section allows users to observe collisions in different reference frames that are related by Lorentz transformations or by Gallilean transformations.

LASER (Lasers), written by Michael Moloney, simulates a three-level laser, with the user in control of energy level parameters, temperature, pump power, and end mirror transmission. Atomic populations may be graphically tracked from thermal equilibrium through the lasing threshold. A mirror cavity simulation is available which uses ray tracing. This permits study of cavity stability as a function of mirror shape and position, as well as beam shape characteristics within the cavity.

HATOM (Hydrogenic Atoms), written by John Hiller, computes eigenfunctions and eigenenergies for hydrogen, hydrogenic atoms, and single-electron diatomic ions. Hydrogenic atoms may be exposed to uniform electric and magnetic fields. Spin interactions are not included. The magnetic interaction used is the quadratic Zeeman term; in the absence of spin-orbit coupling, the linear term adds only a trivial energy shift. The unperturbed hydrogenic eigenfunctions are computed directly from the known solutions. When external fields are included, approximate results are obtained from basis-function expansions or from Lanczos diagonalization. In the diatomic case, an effective nuclear potential is recorded for use in calculation of the nuclear binding energy.

NUCLEAR AND PARTICLE PHYSICS PROGRAMS

NUCLEAR (Nuclear Energetics and Counting), written by Michael Moloney, is included here, but is described under the Modern Physics heading.

SHELLMOD (Nuclear Models), written by Roberta Bigelow, calculates energy levels for spherical and deformed nuclei using the single particle shell model. You can explore how the nuclear potential shape, the spin-orbit interaction, and deformation affect both the order and spacing of nuclear energy levels. In addition, you will learn how to predict spin and parity for single particle states.

NUCRAD (Interaction of Radiation With Matter), written by Roberta Bigelow, is a simulation of alpha particles, muons, electrons, or photons interacting with matter. You will develop an understanding of how ranges, energy losses, and random particle paths depend on materials, radiation, and incident energy. As a specific application, you can explore photon and electron interactions in a sodium iodide crystal which determines the energy response of a radiation detector.

ELSCATT (Electron-Nucleus Scattering), by John Philpott, is an interactive software tool that demonstrates various aspects of electron scattering from nuclei. Specific features include the relativistic kinematics of electron scattering, densities and form factors for elastic and inelastic scattering, and the nuclear Coulomb response. The simulation illustrates how detailed nuclear structure information can be obtained from electron scattering measurements.

TWOBODY (Two-Nucleon Interactions), by John Philpott, is an interactive software tool that illuminates many features of the two-nucleon problem. Bound state wavefunctions and properties can be calculated for a variety of interactions that may include non-central parts. Phase shifts and cross sections for pp, pn, and nn scattering can be calculated and compared with those obtained experimentally. Spin-polarization features of the cross sections can be extensively investigated. The simulation demonstrates the richness of the two-nucleon data and its relation to the underlying nucleon-nucleon interaction.

RELKIN (Relativistic Kinematics), by Joseph Rothberg, is an interactive program to permit you to explore the relativistic kinematics of scattering reactions and two-body particle decays. You may choose from among a large number of initial and final states. The initial momentum of the beam particle and the center of mass angle of a secondary can also be specified. The program displays the final state vector momenta in both the lab system and center of mass system along with numerical values of the most important kinematic quantities. The program may be run in a Monte Carlo mode, displaying a scatter plot and histogram of selected variables. The particle data base may be modified by the user and additional reactions and decay modes may be added.

DETSIM (Particle Detector Simulation), by Joseph Rothberg, is an interactive tool to allow you to explore methods of determining parameters of a decaying particle or scattering reaction. The program simulates the response of high-energy particle detectors to the final-state particles from scattering or decays. The detector size and location may be specified by the user as well as its energy and spatial resolution. If the program is run in a Monte Carlo mode, detector hit information for each event is written to a file. This file can be read by a small reconstruction and plotting program. You can easily modify one of the example reconstruction programs that are provided to determine the mass, momentum, and other properties of the initial particle or state.

QUANTUM MECHANICS PROGRAMS

BOUND1D (Bound States in One Dimension), written by Ian Johnston, is a tool which allows you to explore energy eigenfunctions for an electron in various potential wells, which can be square, parabolic, ramped, asymmetric, double, or Coulombic. The first part of the program deals with finding the eigenvalues and eigenfunctions of different wells. You may find them yourself, using a "hunt and shoot" method, or else the program will compute the eigenvalues automatically, by counting the number of nodes to determine where the eigenvalues occur. The second part of the program looks at properties of eigenfunctions normalization, orthogonality, and the evaluation of many kinds of overlap integrals. The third part examines the time development of general states made up of a superposition of bound state eigenfunctions. Facility is provided for you to incorporate your own procedures to specify different potential wells or different overlap integrals.

SCATTR1D (Scattering in One Dimension), written by John Hiller, solves the time-independent Schrödinger equation for stationary scattering states in one-dimensional potentials. The wavefunction is displayed in a variety of ways, and the transmission and reflection probabilities are computed. The probabilities may be displayed as functions of energy. The computations are done by numerically integrating the Schrödinger equation from the region of the transmitted wave, where the wavefunction is known up to some overall normalization and phase, to the region of the incident wave. There the reflected and incident waves are separated. The potential is assumed to be zero in the incident region and constant in the transmitted region.

QMTIME (Quantum Mechanical Time Development), written by Daniel Styer, simulates quantal time development in one dimension. A variety of initial wave packets (Gaussian, Lorentzian, etc.) can evolve in time under the influence of a variety of potential energy functions (step, ramp, square well, harmonic oscillator, etc.) with or without an external driving force. A novel visualization technique simultaneously displays the magnitude and phase of complex-valued wave functions. Either position-space or momentum-space wave functions, or both, can be shown. The program is particularly effective in demonstrating the classical limit of quantum mechanics.

LATCE1D (Wavefunctions on a one-dimensional Lattice), written by Ian Johnston, is a tool which allows you to explore energy eigenfunctions for an electron in a lattice made up of a number of simple potential wells (up to twelve), which can be square, parabolic, or Coulombic. You may find the eigenvalues yourself, using a "hunt and shoot" method, or allow the program to compute them automatically. You can firstly explore regular lattices, where all wells are the same and spaced at regular intervals. These will demonstrate many of the properties of regular crystals, particularly the existence of energy bands. Secondly you can change the width, depth or spacing of any of the wells, which will mimic the effect of impurities or other irregularities in a crystal. Lastly you can apply an external electric across the lattice. Facility is provided for you to incorporate your own procedures to calculate wells, lattice arrangements or external fields of their own choosing.

BOUND3D (Bound States in Three Dimensions), written by Ian Johnston, is a tool which allows you to explore energy eigenfunctions for a particle in a spherically symmetric potential well, which can be square, parabolic, Coulombic, or several other shapes of importance in molecular or nuclear applications. The first part of the program deals with finding the eigenvalues and eigenfunctions of different wells, assuming that the angular part of the wavefunctions are spherical harmonics. You may find them yourself for a given angular momentum quantum number using a "hunt and shoot" method, or else the program will compute the eigenvalues automatically, by counting the number of nodes to determine where the eigenvalues occur. The second part of the program looks at properties of eigenfunctions normalization, orthogonality, and the evaluation of many kinds of overlap integrals. Facility is provided for you to incorporate your own procedures to specify different potential wells or different overlap integrals.

IDENT (Identical Particles in Quantum Mechanics), written by Daniel Styer, shows the probability density associated with the symmetrized, antisymmetrized, or nonsymmetrized wave functions of two noninteracting particles moving in a one-dimensional infinite square well. It is particularly valuable for demonstrating the effective interaction of noninteracting identical particles due to interchange symmetry requirements.

SCATTR3D (Scattering in Three Dimensions), written by John Hiller, performs a partial-wave analysis of scattering from a spherically symmetric potential. Radial and 3-D wavefunctions are displayed, as are phase shifts, and differential and total cross sections. The analysis employs an expansion in the natural angular momentum basis for the scattering wavefunction. The radial wavefunctions are computed numerically; outside the region where the potential is important they reduce to a linear combination of Bessel functions which asymptotically differs from the free radial wavefunction by only a phase. Knowledge of these phase shifts for the dominant values of angular momentum is used to approximate the cross sections.

CYLSYM (Cylindrically Symmetric Potentials), written by John Hiller, solves the time-independent Schrödinger equation Hu=Eu in the case of a cylindrically symmetric potential for the lowest state of a chosen parity and magnetic quantum number. The method of solution is based on evolution in imaginary time, which converges to the state of the lowest energy that has the symmetry of the initial guess. The Alternating Direction Implicit method is used to solve a diffusion equation given by $HU = -\hbar \partial U/\partial t$, where H is the Hamiltonian that appears in the Schrödinger equation. At large times, U is nearly proportional to the lowest eigenfunction of H, and the expectation value $\langle H \rangle = \langle U|H|U \rangle / \langle U|U \rangle$ is an estimate for the associated eigenenergy.

SOLID STATE PHYSICS

LATCE1D (Wavefunctions for a one-dimensional Lattice), written by Ian Johnston, and included here, is described under the Quantum Mechanics heading.

SOLIDLAB (Build Your Own Solid State Devices), written by Steven Spicklemire, is a simulation of a semiconductor device. The device can be "drawn" by the user, and the characteristics of the device adjusted by the user during the simulation. The user can see how charge density, current density, and electric potential vary throughout the device during its operation.

LCAOWORK (Wavefunctions in the LCAO Approximation), written by Steven Spicklemire, is a simulation of the interaction of 2-D atoms within small atomic clusters. The atoms can be adjusted and moved around while their quantum mechanical wavefunctions are calculated in real time. The student can investigate the dependence of various properties of these atomic clusters on the properties of individual atoms, and the geometric arrangement of the atoms within the cluster.

PHONON (Phonons and Density of States), written by Graham Keeler, calculates and displays phonon dispersion curves and the density of states for a number of different 3-D cubic crystal structures. The displays of the dispersion curves show realistic curves and allow the user to study the effect of changing the interatomic forces between nearest and further neighbor atoms and, for diatomic crystal structures, changing the ratio of the atomic masses. The density of states calculation shows how the complex shapes of real densities of states are built up from simpler distributions for each mode of polarization, and enables the user to match the features of the distribution to corresponding features on the dispersion curves. In order to help with visualization of the crystal lattices involved, the program also shows 3-D projections of the different crystal structures.

SPHEAT (Calculation of Specific Heat), written by Graham Keeler, calculates and displays the temperature variation of the lattice specific heat for a number of different theoretical models, including the Einstein model and the Debye model. It also makes the calculation for a computer simulation of a realistic density of states, in which the user can vary the important parameters of the crystal, including those affecting the density of states. The program can display the results for a small region near the origin, and as a T-cubed plot to enable the user to investigate the low temperature limit of the specific heat, or in the form of the equivalent Debye temperature to enhance a study of the deviations from the Debye model. The Schottky specific heat anomaly can also be investigated.

BANDS (Energy Bands), written by Roger Rollins, calculates and displays, for easy comparison, the energy dispersion curves and corresponding wavefunctions for an electron in a 1-D symmetric $V(x) = V(-x)$ periodic potential of arbitrary shape and of strength V_0. The method used is based on an exact, non-perturbative approach so that the energy dispersion curves and band gaps can be obtained for large V_0. Wavefunctions can be displayed, and compared with one another, by clicking the mouse on the desired states on the energy dispersion curve. Changes in band structure can be followed as changes are made in the shape of the potential. The variation of the band gaps with V_0 is calculated and compared with the two opposite limits of very weak V_0 (perturbation method) and very strong V_0 (isolated atom). Even the experienced condensed matter researcher may be surprised by some of the results! Open-ended class discussions can result from the interesting physics found in these conceptually simple model calculations.

PACKET (Electron Wavepacket in a 1-D Lattice), written by Roger Rollins, shows a live animation, calculated in real time, demonstrating how an electron wavepacket in a metal or semiconducting crystal moves under the influence of external forces. The time-dependent Schrödinger equation is solved in a tight binding approximation, including the external force terms, and the motion of the wavepacket is obtained directly. The main objective of the simulation is to show that an electron wavepacket formed from states with energies near the top of an energy band is accelerated in a direction *opposite* to the direction of the external force; it has a *negative* effective mass! The simulation deals with motion in a 1-D lattice but the concepts are applicable to the full 3-D motion of an electron in a real crystal. Numerical experiments on the motion of the packet explore interesting physics questions such as: how does constant applied force affect the periodic motion of a packet? when does the usual semiclassical model fail? what happens to the dynamics of the packet when placed in a superlattice with lattice constant twice that of the original lattice?

THERMAL AND STATISTICAL PHYSICS PROGRAMS

ENGDRV, written by Lynna Spornick, is a driver program for **ENGINE, DIESEL, OTTO, and WANKEL**. These programs provide an introduction to the thermodynamics of engines.

ENGINE (Design Your Own Engine), written by Lynna Spornick, lets the user design an engine by specifying the processes (adiabatic, isobaric, isochoric [constant volume], and isothermic) in the engine's cycle, the engine type (reversible or irreversible), and the gas type (helium, argon, nitrogen, or steam). The thermodynamic properties (heat exchanged, work done, and change in internal energy) for each process and the engine's efficiency are computed.

DIESEL, OTTO, and WANKEL, written by Lynna Spornick, provide animations of each of these types of engine. Plots of the temperature versus entropy and the pressure versus volume for the cycles are shown with the engine's current thermodynamic conditions indicated.

PROBDRV, written by Lynna Spornick, is a driver program for **GALTON, POISEXP, TWOD, KAC, and STADIUM**. Subprograms GALTON, POISEXP, and TWOD provide an introduction to probability and subprograms KAC and STADIUM provide an introduction to statistics.

GALTON (A Galton Board), written by Lynna Spornick, models either a traditional Galton Board or a customized Galton Board with traps, reflecting, and/or absorbing walls. GALTON demonstrates the binominal and normal distributions, the laws of probability, and the central limit theorem.

POISEXP (Poisson Probability Distribution in Nuclear Decay), written by Lynna Spornick, uses the decay of radioactive atoms to describe the Poisson and the exponential distributions.

TWOD (2-D Random Walk), written by Lynna Spornick, models a random walk in two dimensions. A "drunk," taking equal-length steps, is required to walk either on a grid or on a plane. TWOD demonstrates the joint probability of two independent processes, the binominal distribution, and the Rayleigh distribution.

KAC (A Kac Ring), written by Lynna Spornick, uses a Kac ring to demonstrate that large mechanical systems, whose equations of motion are solvable and which obey time reversal and have a Poincaré cycle, can also be described by statistical models.

STADIUM (The Stadium Model), written by Lynna Spornick, uses a stadium model to demonstrate that there exist mechanical systems whose equations of motion are solvable but whose motion is not predictable because of the system's chaotic nature.

ISING (Ising Model in One and Two Dimensions), written by Harvey Gould, allows the user to explore the static and dynamic properties of the 1- and 2-D Ising model using four different Monte Carlo algorithms and three different ensembles. The choice of the Metropolis algorithm allows the user to study the Ising model at constant temperature and external magnetic field. The orientation of the spins is shown on the screen as well as the evolution of the total energy or magnetization. The mean energy, magnetization, heat capacity, and susceptibility are monitored as a function of the number of configurations that are sampled. Other computed quantities include the equilibrium-averaged energy and magnetization autocorrelation functions and the energy histogram. Important physical concepts that can be studied with the aid of the program include the Boltzmann probability, the qualitative behavior of systems near critical points, critical exponents, the renormalization group, and critical slowing down. Other algorithms that can be chosen by the user correspond to spin exchange dynamics (constant magnetization), constant energy (the demon algorithm), and single cluster Wolff dynamics. The latter is particularly useful for generating equilibrium configurations at the critical point.

MANYPART (Many Particle Molecular Dynamics), written by Harvey Gould, allows the user to simulate a dense gas, liquid, or solid in two dimensions using either molecular dynamics (constant energy, constant volume) or Monte Carlo (constant temperature, constant volume) methods. Both hard disks and the Lennard-Jones interaction can be chosen. The trajectories of the particles are shown as the system evolves. Physical quantities of interest that are monitored include the pressure, temperature, heat capacity, mean square displacement, distribution of the speeds and velocities, and the pair correlation function. Important physical concepts that can be studied with the aid of the program include the Maxwell-Boltzmann probability distribution, fluctuations, equation of state, correlations, and the importance of chaotic mixing.

FLUIDS (Thermodynamics of Fluids), written by Jan Tobochnik, allows the user to explore the fluid (gas and liquid) phase diagrams for the van der Waals model and water. The user chooses four phase diagrams from among the following choices: PT, Pv, vT, uT, sT, uv, and sv, where P is the pressure, T is the temperature, v is the specific volume, S is the specific entropy, and u is the specific internal energy. The program reads in the coexistence table for the van der Waals model and water, and uses it along with an empirical formula for the water free energy and the free energy derived from the van der Waals model. Given v and u, any thermodynamic quantity can be calculated. For the van der waals model thermodynamic quantities also can be calculated from the other thermodynamic state variables. The user can draw a straight line path in one phase diagram and see how this path looks in the other phase diagrams. The user also can extract all important thermodynamic data at any point in a phase diagram.

QMGAS1 (Quantum Mechanical Gas—Part 1), written by Jan Tobochnik, does the numerical calculations necessary to solve for the thermodynamic properties of quantum ideal gases, including photons in blackbody radiation, ideal bosons, phonons in the Debye theory, non-interacting fermions, and the classical limits of these systems. The user chooses the type of statistics (Bose-Einstein, Fermi-Dirac, or Maxwell-Boltzmann), the dimension of space, the form of the dispersion relation (restricted to simple powers), whether or not the particles have a non-zero chemical potential, and whether or not there is a Debye cutoff. The program then allows the user to build up a table of thermodynamic data, including the energy, specific heat, and chemical potential as a function of temperature. This data and various distribution functions and the density of states can be plotted.

QMGAS2 (Quantum Mechanical Gas—Part2), written by Jan Tobochnik, implements a Monte Carlo simulation of a finite number of quantum particles fluctuating between various states in a finite k-space (k is the wavevector). The program orders the possible energy states into an energy level diagram and then allows particles to move from one state to another according to the usual Boltzman probability distribution. Bosons are restricted so that they may not pass through each other on the energy level diagram; fermions are further restricted so that no two fermions may be in the same state; classical particles have no restrictions. In this way indistinguishability is correctly implemented for bosons and fermions. The user chooses the type of particle, the number of particles, the size and dimension of k-space, and the temperature. During the simulation the user sees a representation of the state occupancy and plots of the average energy, the instantaneous energy, and the distribution of energy among the states, also shown are results for the average energy, specific heat, and the occupancy of the ground state.

WAVES AND OPTICS PROGRAMS

DIFFRACT (Interference and Diffraction), by Robin Giles, simulates some of the fundamental wave behaviors in Fresnel and Fraunhofer Diffraction, and other Interference and Coherence effects. In particular you will be able to study diffraction phenomena associated with a point or a set of points and a slit or set of slits using the Huyghens construction. You can also use a method developed by Cornu—the Cornu Spiral—to examine diffraction from one or two slits or one or two obstacles. You can study Fresnel and Fraunhofer diffraction with a single slit or set of slits, a rectangular aperture and a circular aperture. Finally you can study Partial Coherence and fringe visibility in interference and diffraction observations. In the latter example you will be able to study the Michelson Stellar Interferometer and measure the separation distance in a double star and measure the diameter of single stars.

SPECTRUM (Applications of Interfence and Diffraction), by Robin Giles, simulates the uses and modes of operation of four important optical instruments—the Diffraction Grating, the Prism Spectrometer, the Michelson Interferometer and the Fabry-Perot Interferometer. You will look at the nature of the spectra, simulated interference patterns, and the question of resolving power.

WAVE (One-Dimensional Waves), by Wolfgang Christian, Andrew Antonelli, and Susan Fischer, uses finite difference methods to study the time evolution of the following partial differential equations: classical wave, Schrödinger, diffusion, Klein-Gordon, sine-Gordon, phi four, and double sine-Gordon. The user may vary the initial function and boundary conditions. Unique features of the program include mouse-driven drawing tools that enable the user to create sources, segments, and detectors anywhere inside the medium. Double-clicking on a segment allows the user to edit properties such as index of refraction or potential in order to model barrier problems such as thin film interference filters or the Ramsauer-Townsend effect in optics and quantum mechanics, respectively. Various types of analysis can be performed, including detector value, space-time, Fourier analysis and energy density.

CHAIN (One-Dimensional Lattice of Coupled Oscillators), by Wolfgang Christian, Andrew Antonelli, and Susan Fischer, allows the user to examine the time evolution of a 1-D lattice of coupled oscillators. These oscillators are represented on screen as a chain of masses undergoing vertical displacement. The program allows the user to examine how the application of Newtonian mechanics to these masses leads to traveling and standing waves. The relationship between the lattice spacing and other properties such as dispersion, band gaps, and cut-off frequency can be examined. Each mass can be assigned linear, quadratic, and cubic nearest neighbor interactions as well as a time-dependent external force function. Global properties such as the total energy in the lattice or the Fourier transform of the lattice can be displayed as well as the time evolution of a single mass's dynamical variables.

FOURIER (Fourier Analysis and Synthesis), written by Brian James, allows investigation of Fourier analysis and 1-D and 2-D Fourier transforms. In Fourier analysis users can choose from several predefined functions or enter their own functions either algebraically, numerically, or graphically. The build-up of a periodic function is illustrated as successive terms of the Fourier series are added in, and the effects of dispersion and attenuation on the evolution of the synthesized waveform can then be investigated. One- and two-dimensional discrete Fourier transforms can be produced for a range of standard and user-entered functions. The effects of filters on the inverse transforms are illustrated. The 2-D transforms are shown as surface and contour plots. Image processing can be illustrated by filtering the transforms of gray level images so that when the inverse transforms are displayed it can be seen that the images have been modified.

RAYTRACE (Ray Tracing and Lenses), by Brian James, lets the user explore the applications of ray tracing in geometrical optics. The fundamental principle of Fermat can be illustrated by plotting the path of a ray through two different materials between fixed points. The variation of the path of a ray through a region of changing refractive index can be used to investigate the formation of mirages. The variation of pulse delay in a fiber can be calculated as a function of its parameters and the characteristics of optical communication fibers are considered. The formation of primary and secondary rainbows due to dispersion of refractive index can be displayed. The matrix method of tracing rays through lenses can be used to investigate the images formed and show how aberrations in images arise and may be reduced.

QUICKRAY (Quick Ray Tracing), by John Philpott, can be used to demonstrate ray diagrams for a single thin lens or spherical mirror. The object and image are shown, along with the three principal rays that proceed from the object towards the observer. You can use the mouse to move the object, the position of the lens or mirror or to change the focal length of the lens or mirror. The principal rays are continuously redrawn while any of these adjustments are made. The simulation handles converging and diverging lenses and concave and convex mirrors. Thus students can quickly get an intuitive feel for real and virtual image formation under a variety of circumstances.

Acknowledgments

The CUPS Project was funded by the National Science Foundation (under grant number PHY-9014548), and it has received support from the IBM Corporation, the Apple Corporation, and George Mason University.

2

One-Dimensional Bound States

Ian D. Johnston

2.1 Introduction

At the heart of quantum mechanics is the Schrödinger equation, a complex partial differential equation, first order in time, and second order in space. To explore what quantum mechanics says about the world, this equation has to be solved. This is often a difficult exercise. However, there is a whole sub-class of problems which yield valuable insights, and which are less demanding—the so-called **stationary** problems. These have fixed energy and physical properties which do not change in time. A lot can be learned about the "motion" of an atomic particle from them. It involves solving a second-order, real, ordinary differential equation, the **time-independent Schrödinger equation;** the scientific literature of the last half-century is filled with attempts to solve this equation. In general it is necessary to work in three dimensions, but many insights into this more general behavior can be gained from a study of the mathematically simpler, one-dimensional problem.

A discussion of the physical interpretation of the Schrödinger equation and the wave functions which are its solutions can be found in any good textbook on the subject (for example, Merzbacher[1]). What follows is a very brief overview of the most important results that are needed to work with the accompanying program. It will cover a number of quite distinct topics. When you begin work on each topic, you will need to compare the brief description given here with what is in your usual textbook. Make sure at least that you can relate the notation we use to what you are used to.

2.2 The Time-Independent Schrödinger Equation

2.2.1 Introduction

The **time-dependent Schrödinger equation** relates the rate at which a quantum system changes to the total energy of the system (or more accurately, to the Hamiltonian operator); thus,

$$i\hbar \frac{\partial}{\partial t} \Psi(\mathbf{r}, t) = \left(-\frac{\hbar^2}{2m} \nabla^2 + V(\mathbf{r}) \right) \Psi(\mathbf{r}, t) \equiv \hat{H}\Psi(\mathbf{r}, t) \,. \tag{2.1}$$

We usually assume that the potential V does not depend on time. With this assumption this equation has a particular set of solutions which are (relatively) easy to analyze and can yield great insights. These have the same characteristics as the standing waves in a musical instrument. They correspond not to a traveling wave, but to a *vibration*, in which a fixed spatial shape simply fluctuates in amplitude with time. In other words, the time and spatial variations can be separated:

$$\Psi(\mathbf{r}, t) = f(t)\psi(\mathbf{r}) \,. \tag{2.2}$$

Because of the form of Eq. 2.1, this assumption leads directly to the result that such standing waves oscillate harmonically with a frequency related to the total energy of the system,

$$\Psi(\mathbf{r}, t) = e^{\pm iEt/\hbar}\psi(\mathbf{r}) \,, \tag{2.3}$$

and the spatial part $\psi(r)$ satisfies the following equation:

$$\hat{H}\psi(\mathbf{r}) = E\psi(\mathbf{r}) \,. \tag{2.4}$$

The quantity E, which first appears in Eq. 2.1, came about as a constant of integration when the time-varying equation was integrated. What Eq. 2.4 shows is that it is indeed the total energy of the system. Therefore these particular quantum states are states of well-defined energy. In the same way that a study of the acoustical modes of vibration can give information about the sounds a musical instrument can generate, whether steady state or transient, so also can the solutions of Eq. 2.4 give important information about the general behavior of a quantum system. This equation is referred to as the **time-independent**, or **stationary**, **Schrödinger equation.**

Though all real quantum systems (and the Schrödinger equation which describes them) are three-dimensional, there are practical situations which behave approximately as if they were one-dimensional. There are also many insights to be gained from making the simplifying assumption that everything happens in one dimension. That is what we will do in the rest of this chapter.

2.2.2 Bound and Unbound States

The simplest possible situation to which this approach can be applied is that of a particle moving under no external influences, i.e., in a region of zero potential. In that case the Schrödinger equation is

$$-\frac{\hbar^2}{2m}\frac{d^2\psi(x)}{dx^2} = E\psi(x). \tag{2.5}$$

The solutions of this equation can be written either as

$$\psi(x) = A\,\sin(kx) + B\,\cos(kx), \tag{2.6}$$

or, more conveniently,

$$\psi(x) = Ae^{+ikx} + Be^{-ikx}. \tag{2.7}$$

For either of these forms, the wave number k is related to the energy E by

$$k^2 \equiv \frac{2mE}{\hbar^2}, \tag{2.8}$$

and the integration constants A and B will be determined by the boundary conditions which physical requirements impose. These are

- ψ is continuous everywhere,

- $d\psi/dx$ is continuous everywhere, and

- ψ is square integrable.

These are known as **free particle states**. It will be noted that if the quantity k is not real, the solution will become infinite at either $x = +\infty$ or $x = -\infty$. For reasons which will be explained more fully later, it is not considered physically acceptable for a wave function to diverge, and therefore k must always be real for a free particle. This is perfectly reasonable. It says that, if the only energy the particle has is kinetic, the total energy cannot be negative. But there are no other restrictions on the energy, and therefore there are an infinite number of continually variable energy states available to the particle. It is said to have a **continuous spectrum**.

On the other hand, there are many interesting systems in which the particle sees a non-zero potential energy in some limited region of the space. In classical terms, it experiences a force somewhere in this region, wherever the potential is changing. Then the Schrödinger equation becomes

$$-\frac{\hbar^2}{2m}\frac{d^2\psi(x)}{dx^2} + V(x)\psi(x) = E\psi(x). \tag{2.9}$$

The solutions of this equation are no longer as simple as Eq. 2.6 or Eq. 2.7, and indeed finding these solutions is the main job of the accompanying program **Bound1d**.

Note that if the potential V is zero everywhere except in a localized region (where it is positive), then the solutions will have the same asymptotic form as Eq. 2.6. Therefore the particle will still have a continuous spectrum. If, however, the potential is anywhere negative (implying there is an *attractive* force somewhere) then there exists the possibility that there might be solutions with *negative* total energy, and the particle will be largely confined to that region. Such states are called **bound states** (as opposed to **unbound states**, which have positive total energy).

For bound states the asymptotic behavior is most important. If we assume that the potential is truly localized, i.e., that V is different from zero *only* in a limited range of values of x and zero outside that range, then the solution of the Schrödinger equation in the outside regions must be of the form

$$\psi(x) = Ae^{+\alpha x} + Be^{-\alpha x}, \tag{2.10}$$

where

$$\alpha^2 \equiv \frac{2m|E|}{\hbar^2}. \tag{2.11}$$

Again the condition that the wave function should not become infinite at large distances must be invoked. For large negative values of x, the solution would have to have the form

$$\psi(x) = Ae^{+\alpha x}, \tag{2.12}$$

and for large positive values of x,

$$\psi(x) = Be^{-\alpha x}. \tag{2.13}$$

These boundary conditions are very restrictive, and for any general Schrödinger equation there are only a limited number of discrete values of E for which the solution will satisfy them. The physical system is said to have a **discrete spectrum** of bound energy states. (Such a system would *also* be expected to have positive energy solutions: its *complete* spectrum would have a continuous part and a discrete part.)

The mathematical study of systems like this is a well-studied branch of mathematics called Stürm-Liouville theory. The technical term **eigenvalue** was invented for these special discrete values of the parameter for which a solution exists which satisfies the boundary conditions. Its use was later extended to apply to cases where the set of such values is continuous. The solution corresponding to any eigenvalue is called an **eigenfunction**. And in quantum mechanics the physical state described by such a solution is known as an **eigenstate**.

In the current chapter we devote attention to discrete, bound eigenstates.

2.2.3   Bound Eigenstates

Since we will only be considering systems whose total energy is negative, it is convenient to introduce a positive definite quantity defined by

$$E_B \equiv -E. \tag{2.14}$$

For eigenstates this quantity is known as the **binding energy**. With this definition the Schrödinger equation can be rewritten as

$$\frac{d^2\psi(x)}{dx^2} = \frac{2m}{\hbar^2}(E_B + V(x))\psi(x). \tag{2.15}$$

It is important to be aware of the units of this equation. If x is chosen to be measured in an arbitrary-length unit which we will call L_0, and all energies in another arbitrary unit E_0, then Eq. 2.15 can be written in a completely dimensionless form,

$$\frac{d^2\psi(x)}{dx^2} = C(E_B + V(x))\psi(x), \tag{2.16}$$

where x, E_B, and V are all dimensionless quantities and the constant C is defined by

$$C \equiv \frac{2mE_0L_0^2}{\hbar^2}. \tag{2.17}$$

Many advanced textbooks on quantum mechanics choose to work in units which effectively make this constant equal to 1. This not only means a great simplification of the algebra but it also makes the important point that quantum behavior, as exemplified by the Schrödinger equation, is similar for many systems. Problems dealing with electrons in atomic potentials, or nucleons in nuclear potentials, for example, can be directly related to one another simply by changing the value of the measurement units L_0 and E_0.

In this chapter and its accompanying program, however, we will consider exclusively *electrons* moving in potentials of roughly atomic dimensions. All lengths will be measured in **nm**, and all energies in **eV**. However, at some stage you may wish to change the program to work with, say, fm and MeV. If so, you should change the part of the program which calculates the program variable **energyConversion**, which is the same as the C in Eq. 2.16 and Eq. 2.17.

2.2.4 Properties of Wave Functions

There are several general properties that all sets of eigenfunctions possess, and learning to understand and use these is one of the main jobs facing any student of the subject.

In what follows we will use a single index n (a **quantum number**) to identify different eigenfunctions belonging to the set. Any one eigenfunction is conventionally written as $u_n(x)$ (to distinguish it from a more general state of the system usually denoted by $\psi(x)$). The corresponding energy eigenvalue is written as E_n. (The binding energy, which will be written E_{Bn}, is of course the negative of this.)

- **Normalization**

  The standard interpretation of quantum mechanics is that the magnitude of the wave function is related to the probable outcome of an experiment which aims to measure where the particle is. Specifically, if the probability of finding the particle between the points x and $x + \Delta x$ at time t is written $P(x, t)\Delta x$ (where P is called the **probability density**), then

  $$P(x, t) = \Psi^*(x, t)\Psi(x, t). \tag{2.18}$$

  This provides a commonsense way of specifying the amplitude of the spatial part of the wave function. If there is only one particle in the system, the total probability of finding the particle *somewhere* should be unity. This means

  $$\int_{-\infty}^{+\infty} dx\,\psi^*(x)\psi(x) = 1. \tag{2.19}$$

  The process of adjusting the amplitude of the wave function so that Eq. 2.19 is true is known as **normalization**.

Note that the physical ideas involved here provide the explanation for the requirements imposed earlier on the wave function. Firstly, the integral in Eq. 2.19
must exist, which is why we said the wave function had to be square integrable.
Secondly, the wave function should never be allowed to become infinite at great
distances because, if it did, it could not be interpreted as describing a particle
that was confined locally. So this interpretation implies that, unless the energy
of the system happens to match one of the eigenvalues exactly, the particle will
not be found within the well so long as we require that its wave function oscillates harmonically with time. It may move in and out of the well, but it cannot
stay there.

- **Orthogonality**
  A purely mathematical property of any set of *eigenfunctions* is the following.
  If any two different eigenfunctions are multiplied together (taking the complex
  conjugate of one of them), and the product integrated from $-\infty$ to $+\infty$, the integral will vanish:

$$\int_{-\infty}^{+\infty} dx\, u_n^*(x)u_m(x) = 0 \quad \text{where} \quad m \neq n. \tag{2.20}$$

Note that, if the eigenfunctions had previously been normalized, as we will assume they have been, this result can be written using the Kronecker delta; thus,

$$\int_{-\infty}^{+\infty} dx\, u_n^*(x)u_m(x) = \delta_{nm}. \tag{2.21}$$

- **Expectation Values**
  A wave function contains more information than just *where* the particle is. Any
  dynamical property—position, momentum, energy, etc.—can be "measured"
  by calculating what is the probable outcome of an experiment to find the value
  of one of these quantities. The result of this calculation is known as the **expectation value**, and it is found by using the following prescription:

$$\langle A \rangle = \int_{-\infty}^{+\infty} dx\, \psi^*(x)\hat{A}\psi(x). \tag{2.22}$$

The **operator** $\hat{A}$ for any classical dynamical variable which is a function of x
and p ($A(x, p)$) is constructed by making these replacements (with due respect
for order)

$$x \longrightarrow x \quad \text{and} \quad p \longrightarrow \frac{\hbar}{i}\frac{\partial}{\partial x}. \tag{2.23}$$

Eq. 2.22 assumes that the wave function ψ has been normalized. If this is not
the case, then the right-hand side would have to be divided by the appropriate
value of the normalization integral.

- **Completeness**
  There are many other wave functions, other than eigenfunctions, which satisfy
  the same boundary conditions yet are not solutions of the time-independent
  Schrödinger equation. It is a well-known property, which can be established in

special cases but is assumed in general, that any such function $\psi(x)$ can always be expressed as a *linear combination* of the eigenfunctions:

$$\psi(x) = \sum_n a_n u_n(x) \, . \tag{2.24}$$

This property of the set is known as **completeness**.

For Eq. 2.24 to be valid it is important that *all* the eigenfunctions be included. If some of the eigenvalues form a continuous spectrum, the summation in this equation must be understood to include an *integration* over the continuous quantum numbers, *as well as* a sum over the discrete ones. If the set of eigenfunctions are normalized, and since they are orthogonal to one another, Eq. 2.24 can be inverted to show how the coefficients a_n may be calculated:

$$a_n = \int_{-\infty}^{+\infty} dx \, u_n^*(x)\psi(x) \, . \tag{2.25}$$

To help in the understanding of these ideas, it is useful to think about the simple theory of vibrations. The *Fourier theorem* says that any periodically repetitive shape is equivalent to a linear combination of sinusoids (i.e., the standing waves are a complete set). Any vibration of a one-dimensional medium will therefore be equivalent to a linear combination of sinusoidal modes of oscillation, each vibrating independently, with the amplitude that was given it by the original disturbance.

Quantum mechanics uses exactly analogous ideas. In most real physical situations it is unlikely that the total energy exactly coincides with one of the eigenvalues. Therefore at any instant of time a real system can be expected to be described by a spatial wave function which has the correct boundary conditions (i.e., it goes to zero at $x = \pm\infty$), but it usually does NOT satisfy the stationary Schrödinger equation. There is no generally agreed-upon name for such a state. In this chapter and the program we will refer to it as a **general state**. However, we will always discuss the behavior of such a state by considering it as a superposition of the eigenstates. Furthermore, in Fourier theory, most real vibrations can be approximated by a small number of the low-frequency harmonics. Similarly, we would expect that we would not have to bother much with the continuum states of any particular potential well. Most of the physical behavior we want to study would be shown by linear combinations of the discrete bound eigenstates.

- **Matrix Elements**

  The Schrödinger approach to quantum mechanics is a way of solving problems for quantum systems by treating the basic laws as differential equations and assigning differential operators to physical observables. However, it is just one "representation" of quantum mechanics. There are other ways of tackling the same problems. Another is **matrix mechanics**, invented by Heisenberg in 1927, which represents quantum states by ($N \times 1$) column matrices, and operators by ($N \times N$) square matrices. (Here N is the number of eigenvalues, which may well be infinite.)

The link between the two representations is achieved by the same kind of integrals we have been discussing here. The m,n element of the Heisenberg matrix A is related to the Schrödinger operator $\hat{A}$ by

$$A_{mn} = \int_{-\infty}^{+\infty} dx\, u_m^*(x)\hat{A}u_n(x)\,. \qquad (2.26)$$

As might be expected, since the two representations are solving the same physical problems, these matrix elements occur in many places in both ways of doing quantum mechanics. For example, there is a widely used technique of finding approximate energy eigenvalues known as **perturbation theory**. We will not describe it here: you will find it discussed in any of the textbooks in the bibliography (see, for example, Merzbacher,[1] chapter 17). Suffice it to say that it gives an expression for corrections to energy eigenvalues as the diagonal elements of a Hamiltonian matrix, and corrections to energy eigenfunctions in terms of the off-diagonal elements of the same matrix.

It will be noticed that in order to use any of the properties of wave functions outlined here, you have to be able to calculate the value of an **overlap integral**— where the integrand is the product of two eigenstates and/or general states, with an operator acting on one of them. (For generality the factor 1 counts as an operator too.) The second part of the **Bound1d** program therefore does nothing else but calculate these overlap integrals.

2.2.5   Time Development

The complete wave function for an electron which is in an energy eigenstate is

$$\Psi(x, t) = e^{-iE_n t/\hbar} u_n(x)\,. \qquad (2.27)$$

It will immediately be noticed that the probability density $\Psi^*(x, t)\Psi(x, t)$ is independent of time. This reinforces the conclusion drawn earlier that an energy eigenstate is a *stationary* state.

As previously discussed, in general it is unlikely that a particular physical system will be in an eigenstate, but at some initial time ($t = 0$) can be expanded as a linear combination of eigenstates:

$$\Psi(x, 0) = \sum_n a_n u_n(x)\,. \qquad (2.28)$$

Each of the eigenstates will evolve separately with time. Therefore at some later time the complete wave function will be

$$\Psi(x, t) = \sum_n a_n e^{-iE_n t/\hbar} u_n(x)\,. \qquad (2.29)$$

When these time-dependent states are calculated, it is necessary to choose a sensible unit of time , otherwise everything will happen too fast or too slowly. The natural time scale on which any general state will change appreciably is of the order of $\hbar/E$, where E is a typical energy of the system. Hence, since we will be dealing with energies measured in eV, a natural time unit is

$$t_0 = \frac{\hbar}{1eV} = 6.586 \times 10^{-16} \text{ s}. \tag{2.30}$$

The third part of the program shows how different initial states evolve with time, and in that part of the program time will be measured in terms of this time unit.

2.3 Computational Approach

2.3.1 Numerical Solution of the Schrödinger Equation

Ever since it was conceived in 1927, quantum mechanics, at least in the estimation of non-specialists, has been bedevilled by the mathematical complexity of its formulation. The central problem is to solve the Schrödinger equation (Eq. 2.9), but this is not easy to do in general, owing largely to the uncertainty of the functional form of the potential *V*. The last sixty years have seen many elegant solutions to different problems of this kind, often with a new method for each case. It is sensible therefore to look for a straightforward method of solving Eq. 2.9, which may not be the most elegant nor the most accurate in all cases, but which can usually be relied upon to work.

With the advent of personal computers (where there is plenty of time to experiment) with good graphics (so what is produced can be seen) such an approach is possible. The key idea is to approximate the process of integration, necessary to "solve" Eq. 2.9, as a coarse-grained summation. One simple approximation, very straightforward to verify, is to expand the second derivative thus:

$$\frac{d^2}{dx^2}\psi(x)\Bigg]_{x=x_i} \approx \frac{\psi(x_i + \Delta x) - 2\psi(x_i) + \psi(x_i - \Delta x)}{(\Delta x)^2}. \tag{2.31}$$

Hence, knowing ψ at x_i and $x_{i-1}(= x_i - \Delta x)$ allows it to be calculated at $x_{i+1}(= x_i + \Delta x)$, using Eq. 2.9 and Eq. 2.31 together. The process is continued as long as necessary by simple repetition of the process. Note, however, that two values of ψ were needed to start with.

The process just described goes by the name of the "half-step method." Though it is simple in principle, it often proves a bit too inaccurate, especially for the kind of integrations we want to do here. It is then necessary to use a more accurate method. Such a method goes by the name of its author, Numerov. A complete description can be found in Friar.[2] This method is used in the present program. However, it is important to realize that it is much the same in principle. It starts with two values of the function at neighboring points on the *x*-axis—or a value of the function and its derivative at one point, which is equivalent—and then the function is calculated at the next point, repeating the process as long as necessary.

A good description of the numerical solution of the Schrödinger equation has been given by French and Taylor.[3]

2.3.2 Eigenvalues of Bound State Problems

The method just described will find a "solution" of the Schrödinger equation under essentially any circumstances. However, as pointed out earlier, a physically acceptable wave function must approach zero asymptotically at $x = \pm\infty$. If a value of E_B is selected at random, the Schrödinger equation can be integrated numerically, starting from physically acceptable initial conditions for large negative values of x. However, the chances are very small that at large positive x, the solution will approach the axis asymptotically (as it should). It is much more likely that it will diverge, either up or down, away from the axis. However, there are some values of the E_B for which it diverges neither up nor down, but approaches the axis "just right." The problem facing someone trying to investigate the Schrödinger equation computationally is to develop a way to locate these eigenvalues. Once we have that, we may integrate the equation for just those binding energies and see what the corresponding eigenfunctions look like.

There are many ways to do this. One, not very sophisticated but easy to understand, is called the "hunt-and-shoot" method. It relies on two important properties of the set of eigenfunctions of any Schrödinger equation.

1. Each of the eigenfunctions has a unique number of **nodes** (points at which the function crosses the axis). The ground state has none, the first excited state one, the second two, and so on. A solution of the equation corresponding to value of E_B which is *not* an eigenvalue has the same number of nodes as the eigenfunction with the next highest eigenvalue.

2. If the solution for a general value of E_B starts as a *positive* increasing exponential for large negative values of x, its asymptotic value for large positive values of x will change sign as E_B passes through an eigenvalue: i.e., if it diverges above the x-axis for E_B slightly greater than one eigenvalue, it will diverge below the axis for E_B slightly less than that eigenvalue, and vice versa (see Fig. 2.1).

Finding a solution computationally therefore involves experimenting with the program, seeing what happens when various values of E_B are tried. If a value is found for which the solution diverges upward at large x, and another for which it diverges downward, then somewhere between the two there must be at least one eigenvalue. Then a systematic searching process can be launched (it can be done by hand, though the program does it automatically). Choose a value halfway between the two initial guesses for the binding energy and see what the solution does there. It may diverge above or below the x-axis for large x. Either way, the range in which the eigenvalue lies has been narrowed. Keep doing this until an E_B is found at which it doesn't go up or down, to whatever accuracy the calculation is working. This is known as the **binary search method**. More efficient searches exist and may be found in any good book on numerical analysis (for example, Press et al.[5]).

The process of finding a pair of energy values to start the search process can be very difficult. In many potential wells of interest energy levels are very close together, and relying on finding by chance an energy which separates them can get very tedious. However, because there is a relation between the number of nodes and the nearest eigenvalue, *counting* the number of nodes in any solution can also be used to home in on an eigenvalue. The program has a section which does this automatically.

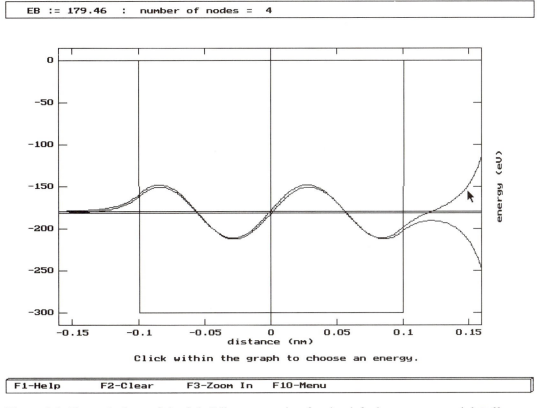

```
EB := 179.46  :  number of nodes =  4
```

Figure 2.1: Two solutions of the Schrödinger equation for the default square potential well—one with $EB = 179.46$ eV, which diverges upward for large positive values of x, and the other with $EB = 181.43$ eV, which diverges downward. The eigenfunction with $n = 4$ lies between these two values and is at $EB = 180.44$ eV.

2.4 Exercises

Many different calculations can be done with the program **Bound1d**. A few are listed here as suggested exercises. Some are actually projects that require modification of the program. Before trying the exercises, review the section on running the program.

2.4.1 Finding Eigenvalues and Eigenfunctions

2.1 **Square Well**

Work with the default square well, whose parameters are depth, 300 eV; width, 0.20 nm. However, before you start, work out analytically the energy levels of an *infinite* well of the same width.

a. Describe what solutions of the wave equation look like for different energies (use the menu item **Method | Try Energy**; i.e., select **Try**

Energy from the pull-down menu **Method**). Try to home in on one eigenfunction.

b. Next do an automated binary search (**Method | Hunt for Zero**), and find the energy levels of the well. Note:

$$\text{with } L = 0.20 \text{ nm}, \quad E = \frac{h^2}{8m_e L^2} \approx 9 \text{ eV}.$$

Search around this value above bottom of well. Search around n^2 times this value.

c. Show the complete eigenvalue spectrum (**Spectrum | Find Eigenvalues**). Examine each eigenfunction (**Spectrum | See Wave Functions**). Comment on the symmetry properties of the wave functions, and the asymptotic values.

2.2   Scaling

It should be clear that if you make the square well deeper (or shallower) the eigenvalues will change. However, you can compensate by making the well narrower (or wider) so that the dimensionless form of the Schrödinger equation remains the same.

a. Derive analytically an expression for this scaling law.

b. Now check that result computationally—i.e., construct another well which is shallower and wider (in the correct ratio), and, before you start, calculate what the four energy levels should be. Then get the computer to measure them. How accurately did it agree with what you predicted?

2.3   Asymmetric Well

Consider the well to be divided into two equal parts, one side being deeper than the other, like the following:

$$V = 0 \qquad \text{for } |x| > r_1/2 \qquad\qquad (2.32)$$
$$= -V_1 \quad \text{for } -r_1/2 \leq x \leq r_2$$
$$= -V_2 \quad \text{for } r_2 < x \leq r_1/2.$$

The default values in the program are reasonable ones to start with: $r_1 = 0.20$ nm; $r_2 = 0.00$ nm; $V_1 = 300$ eV; $V_2 = 100$ eV.

a. Investigate how the energy levels change as the relative depth of the two halves of the well change. Set V_2 to be a fraction f of V_1, and then find all the energy eigenvalues for values of f equal to 0.0, 0.2, 0.4, 0.6, 0.8, and 1.0. Display the results by plotting (by hand is probably easiest) the energy eigenvalues on an energy versus f graph. Can you come up with a physical explanation/description of what this graph shows?

b. Investigate how the shape of the wave functions alter as the relative *width* of the two parts of the well change. Choose the same starting values as in part 1: $r_1 = 0.20$ nm; $r_2 = 0.00$ nm; $V_1 = 300$ eV; $V_2 = 100$ eV.

Make a note of how the peak values of the two halves of the wave function change as the fraction r_2/r_1 goes from -0.5 to 0.5 (in steps of, say, 0.25). You should note that at some points the two amplitudes are nearly equal. As an exercise, try to find the value of the fraction r_2/r_1 where the two parts of the wave function corresponding to $n = 3$ (i.e., the second excited state) have essentially the same amplitude. Can you see the mathematical reason why the two amplitudes are the same? Does this have a physical explanation?

2.4 Double Square Well
This well is defined by

$$V = 0 \qquad \text{for } |x| > r_1/2 \qquad\qquad (2.33)$$
$$= -V_1 \quad \text{for } r_1/2 \le |x| \le r_2/2$$
$$= -V_2 \quad \text{for } 0 < |x| \le r_2/2 .$$

Choose some particular values of the variables to get you going—say, $r_1 = 0.20$ nm; $r_2 = 0.10$ nm; $V_1 = 300$ eV; $V_2 = 100$ eV.

a. Find the first three energy eigenstates and their eigenvalues.

b. Observe the near degeneracy of pairs of states. Explain the reason for this in terms of the near identity of the probability functions.

c. Now make the central barrier height $V_1 - V_2 = fV_1$, and investigate what happens to the wavefunctions and their eigenvalues as the fraction f changes from 0.0 to 1.0, in steps of 0.25. Plot the eigenvalues versus f.

d. Think about the limiting case, if this barrier were infinitely high (i.e., if f were very large). Without doing any computing, what do you think the eigenfunctions and eigenvalues should be? If you make f greater than 1, is this what you get? (Warning: The program will not allow you to enter a negative value of V_2 greater than 100 eV. This is because the energy eigenvalues can get so close together that the program will crash when trying to separate them.)

e. When you come to think about it, it seems more physically intuitive that if the central barrier is so high that the two side wells are absolutely isolated from one another, the electron should be in one well or the other, but not in both. Can you think of a way of reconciling this commonsense view with the straightforward use of the Schrödinger equation we have adopted up till now? (Hint: Think about general properties of the solutions of differential equations.)

2.5   Covalent Bonding

The simplest possible *molecular* structure that quantum mechanics can explore is the singly ionized hydrogen molecule. Even though the calculation of its ground state was one of the early successes of quantum mechanics, even today it remains a difficult analytical problem. It is solved exactly in chapter 8, but at this point you can approximate the solution in one dimension.

Introductory textbooks on atomic physics usually describe the calculation like this. The electron spends half its time orbiting round one nucleus and half about the other. So the total wave function must look somewhat like the sum of two single atomic wave functions, normalized so that the total probability comes to 1:

$$\psi_{total}^{+} = \frac{1}{\sqrt{2}} \times (\psi_1 + \psi_2). \tag{2.34}$$

But, since the sign of a wave function is not normally measurable, the "real" wave function might just as well be their difference:

$$\psi_{total}^{-} = \frac{1}{\sqrt{2}} \times (\psi_1 - \psi_2). \tag{2.35}$$

Subtle experimental evidence suggests that the state ψ^{+} (called the **symmetric state**) is the only one that is ever observed. This result has a special relevance here. Imagine you are constructing the ψ^{+} state by bringing a neutral hydrogen atom and a proton together from infinity. The total energy of the pair must become more and more negative as they come closer together, even allowing for the electrostatic repulsion of the two nuclei, which contributes a positive potential energy. After a while, however, the repulsion must win out, and the total energy will be positive for very close separations. The natural resting place of the two protons will be at the minimum of this potential energy curve. On the other hand, it must also mean that the total energy of the state ψ^{-} (the **anti-symmetric state**) is never negative. That is why no bound state exists for that configuration.

When you looked at the double square well in the previous exercise you should have gotten enough insight to understand much of what those atomic physics textbooks are saying. It ought to be possible to construct a one-dimensional model using simple square wells to explain the phenomenon of covalent bonding. Try to do this, remembering all the time that you can't hope to predict any actual numbers, so you are at liberty to change any well parameters you want.

a. Start off with two identical wells quite close together, and choose parameters so that there is a reasonable separation between the energies of the first two states. It is also important that each individual well should be deep enough to have a ground state that is easy to calculate. You will then need to separate the wells bit by bit so that they approach the state where they are essentially independent. Note the binding energy of the system at each separation for the ground state and the first excited state.

b. At the end plot the total energy of the system—electronic energy *plus* the Coulomb repulsion of the two protons. Is there a point at which the symmetric state has a minimum of the energy? What about the anti-symmetric state?

2.6 **Parabolic Well**

The Schrödinger equation for a particle moving in a harmonic oscillator potential is

$$-\frac{\hbar^2}{2m}\frac{d^2\psi}{dx^2} = \left(E - \frac{1}{2}m_e\omega^2x^2\right)\psi(x). \tag{2.36}$$

a. Write down what you know the eigenvalues of this equation to be.

b. Write down the algebraic form of a potential well which is parabolic in shape, and which goes to V_0 at the center of the well ($r = 0$) and reaches zero at ($x = \pm r_1$). Then write down the Schrödinger equation with this potential included. By comparing this equation with that in part 1, write down—

1. the algebraic relation between the classical frequency ω and the parameters of the well V_0 and r_1; and

2. the relation between the binding energy E_B in Eq. 2.15 (which must be positive, remember) and the energy E in Eq. 2.36 (also positive). What you should get should simply reflect the fact that, when you made the center of the well negative, you shifted the zero for measuring energies.

Note: For the way the well is drawn, the Schrödinger equation is

$$-\frac{\hbar^2}{2m_e}\frac{d^2\psi}{dx^2} + \left(\frac{V_0}{L^2}\right)x^2\psi = (E - V_0)\psi = -\frac{\hbar^2}{2m_e}\frac{d^2\psi}{dx^2} + \frac{1}{2}m_e\omega^2x^2\psi.$$
$$\tag{2.37}$$

If you use the default values $V_0 = 300$ eV and $L = 0.1$ nm, the following can be calculated

$$\omega = 1.12 \times 10^{17} \text{ s}^{-1}, \ T = 5.60 \times 10^{-17} \text{ s}, \ \hbar\omega = 74 \text{ eV},$$

$$\text{length scale} = \sqrt{\frac{m\omega}{\hbar}} = 0.289 \text{ nm}^{-1}$$

$$\text{time scale} = \frac{\hbar}{V_0} = 2.20 \times 10^{-18} \text{ s}.$$

c. Choose a parabolic well from the program menu (**Potential**). Demonstrate firstly that the levels are equally spaced, and then that there is close (but not perfect) agreement with theoretical eigenvalues. This occurs because the theoretical result assumes the potential kept increasing towards $x = \pm\infty$, whereas the computational potential is truncated and set to zero outside the range.

d. Measure positions of nodes of the different eigenfunctions by choosing the menu item **Method | Examine Solution**. Compare the results with the zeroes of the corresponding Hermite polynomials.

2.7   A Perturbing Electric Field

Add a small perturbing term to any of the standard potential wells, say, a weak electric field, and see what difference it makes to the energy eigenvalues.

2.4.2   Properties of Bound State Functions

2.8   Orthonormality

Run part 2 of the program dealing with wave function properties, using the **asymmetric well**, and choosing the menu item **Psi 1 | Eigenstate, Psi 2 | Eigenstate**. If you select the default well parameters and find the eigenfunctions, you should observe that there are five.

a. Select the ground state energy eigenvalue for both input variables. Verify that the program has correctly normalized the ground state eigenfunction. Similarly, verify that the other four eigenfunctions are correctly normalized.

b. Now verify that all five eigenfunctions are orthogonal to one another. (Since there are five different eigenfunctions, there are 25 different overlap integrals that you must look at.) How accurately do these integrals vanish?

2.9   Expectation Values

By choosing the menu item **Operator | x**, get the program to calculate the expectation value of the position of the particle when it is in an eigenstate.

a. Find $\langle x \rangle$ for each of the eigenstates. Would you have gotten the right answer if the wave functions had not been normalized?

b. Similarly calculate $\langle KE \rangle$, $\langle PE \rangle$, and $\langle E \rangle$ for all three eigenstates. Observe carefully the shapes of the functions when they are operated on by these three operators.

2.10   Double Well

Repeat everything you just did in the preceding two questions using the potential well with a central barrier **(Double Well)**. There are a few extra questions you should try to answer in the different parts of the question.

a. When you look at the shape of the eigenfunctions you should be able to tell immediately, even before doing the integration, that there are two pairs which must be orthogonal to one another. Explain why (use arguments of symmetry).

b. Explain why you get the same value of $\langle x \rangle$ for all eigenstates.

c. Repeat the calculations for $\langle PE \rangle$, $\langle KE \rangle$, and $\langle E \rangle$ for different barrier heights. Explain what you observe.

2.11 Momentum Expectation Values

Demonstrate that the expectation value of p is always zero for a bound eigenstate, for any well shape. Why is this so?

2.12 Hermiticity

Demonstrate that x is a hermitian operator, and that $\frac{d}{dx}$ is anti-hermitian, for any potential shape.

2.13 Matrix Elements

Choose the parabolic well and calculate the matrix elements of x and $\frac{d}{dx}$. Compare what you get with the theoretical values for x and p with a harmonic oscillator potential. Note: You should find the explicit answer to the previous question for this potential when you write out all the elements as a matrix.

2.14 Perturbation Theory

For a well of simple shape (say, a square well), use perturbation theory to calculate the first-order change in the energy eigenvalue of any state due to a small applied electric field. Do this by using the following method.

a. Assume the eigenfunctions you have just calculated to be "unperturbed" states. Write down a theoretical expression for the change in energy of any one level, using as the perturbation the potential energy arising from an applied electric field.

b. Now evaluate the overlap integral in this expression by choosing the appropriate operator from the menu.

c. Check the answers you get with the results of Exercise 2.7.

2.15 Electric Polarizability

Using a simple well shape with an applied electric field, calculate the electric dipole moment $(e\langle x \rangle)$. Plot it as a function of E and deduce the polarizability of a single "atom."

2.16 Permanent Dipole Moment

Choose a double well, consisting of two square wells, separated so that the first two eigenvalues are very close together. Then add a small electric field. Calculate the electric dipole moment for this "atom," and its polarizability, in either the ground or first excited state. How does it differ from the behavior of the "atom" in the previous question?

2.17 Operators and Eigenfunctions

Demonstrate that the energy eigenfunctions are unchanged in shape by operation of E operator.

2.18   **Uncertainty**

Choose a square well, and calculate the uncertainty in x and p. How does your answer compare with the Heisenberg uncertainty principle?

2.19   **Commutation Relations**

For any well, calculate matrix elements of $[x, p]$. Verify the theoretical result that

$$[x, p] = i\hbar \,. \tag{2.38}$$

2.20   **General States**

Demonstrate how to normalize a general state (**Psi 1 | General State, Psi 2 | General State**). Suggestion: Form a general state by combining two eigenstates with suitable coefficients.

2.21   **Two-State Systems**

Choose a double square well, separated so that the first two eigenvalues are very close together. Construct a general state in which the electron is localized in either well. (Hint: Choose coefficients equal to $1/\sqrt{2}$.) Calculate $\langle x \rangle$, and the uncertainties in x and p. Calculate $\langle E \rangle$.

Now change the sign of one of the coefficients so that the electron is localized in the other well. Calculate $\langle x \rangle$, and the uncertainties in x and p and $\langle E \rangle$ for this state. Comment.

2.22   **Completeness**

This is an exercise in constructing a general state of some particular shape, using a simplified form of the completeness relation. The specific aim will be to construct a **wave packet**.

   a. Choose the ramped well to form a set of base states. Use the menu item **Psi 2 | Some Other Function**, which will ask you to input a functional form. You can enter a Gaussian shape if you wish, or a simple "hat" function, which is the default. Calculate the overlaps of this function with all the eigenfunctions of the well.

   b. Use the results as coefficients to synthesize a new general state. How well are you able to reproduce your original function? (Remember that to get a complete synthesis you would have to use the continuum states also, which we cannot do here.)

   c. Using the general state you have constructed, first of all normalize it, and then calculate $\langle x \rangle$ and $\langle E \rangle$. Comment on the relation of these two quantities.

2.4.3   Time Development of Bound State Wave Functions

2.23   **Two State Transitions**

Choose the double square well and construct a localized state, as in Exercise 2.21. Demonstrate how the particle moves (**Wave Func | Show Time Development**).

You should observe that the electron moves back and forth between the two wells. How long does it take to do this? How should this time relate to the energies of the two states? Use this as a model for what would happen if an "atom" like this were exposed to an oscillating electric field.

2.24 Oscillating Wave Packet

Choose the default parabolic well. Use the methods of Exercise 2.22 to find the coefficients to construct a wave packet representing the particle pulled to one side with zero momentum (i.e., all real coefficients).

a. Describe how the particle moves. Comment on the fact that the wave function comes back into phase at the extremity of each swing. (Note: The following coefficients give quite a reasonable wave packet (unnormalized),

$$+0.0640, \ -0.1269, \ +0.1476, \ -0.1029, \ +0.0219.)$$

b. Measure $\langle x \rangle$ at various times (**Measure | Position, Measure | Set Time**) and plot it as a function of time. Explain what you see.

2.25 Bouncing Wave Packet

Choose a ramped well that is 300 eV deep on one side and 0 eV on the other (with the default width). Construct a wave packet representing the particle at the top of the ramp at rest.

a. Describe how the particle moves. (Note: The following coefficients give a reasonable wave packet.

$$+0.0276, \ -0.1185, \ +0.0949, \ +0.0644.)$$

b. Again plot $\langle x \rangle$ as a function of time, and explain what you see. Does the wave packet bounce with an abrupt change of velocity like a tennis ball?

2.26 Coherent States

In the quantum theory of radiation particularly important general states are the so-called **coherent states** of the harmonic oscillator (see, for example, Haken[4]). These are defined by

$$\psi(x) = \sum_n \frac{\alpha^n}{\sqrt{n}} u_n(x). \tag{2.39}$$

(Note: This form is not normalized.) Construct one of these states, putting α equal to 1 and using only the seven lowest eigenstates. You should be able to see why this state is considered special as you observe how it develops with time. (Note that, because you are using so few states to synthesize the packet, it may soon start to lose its nice shape.)

2.5   Details of the Program

2.5.1   Running the Program

On first loading the program you are presented with a credit screen which can be cleared by pressing any key or clicking the mouse. You are then presented with the following menu of choices:

File   Parts   Potential   Parameters   Method   Spectrum

Move the highlighted item in the menu by the right and left <**arrow**> keys, and press <**Enter**> when you have reached the one you want. There are three different parts of the program, and this main menu changes depending on which part of the program you are currently working in. The first two menu items are common to all three parts. Choosing either of them has the following effects:

● **File**

  – **About Program**: This gives a brief description of what the program does.

  – **About CUPS**: This gives a brief description of the CUPS program.

  – **Configuration**: This allows you to set a path for the storage of temporary files, to change the colors of the display, and to check how much memory has been used.

  – **Open ...**: This allows you to read in default values from a file you have previously saved from this program. Choose the name of the file from the list presented. If the file you want is not on the disk, press <**Esc**> to exit. The default values will not be changed.

  – **Save**: This allows you to save all the values you have entered into all the input screens of the program, so that you can start where you left off when running the program in future. If you opened a file earlier in the session, the information will be written into that file. If not, you will be asked to supply a name for the file.

  – **Save as ...**: This allows you to save all the values you have entered into all the input screens of the program, so that you can start where you left off when running the program in future. You will be asked to supply a name for the file.

  – **Exit Program**: This takes you out of the whole program.

● **Parts**

  – **About Part 1/2/3**
    This gives a brief description of the part of the program you are currently in.

  – **Part 1: Finding Eigenvalues**
    This puts you into part 1 of the program.

– **Part 2: Wave Function Properties**
This puts you into part 2 of the program.

– **Part 3: Time Development**
This puts you into part 3 of the program.

Part 1: Finding Eigenvalues

This part of the program is used to find the eigenfunctions and eigenvalues of a number of different potential wells. In this part of the program the screen layout becomes that shown in Figure 2.1. The last three menu items have become the following.

● **Potential**
This allows you to choose from the following well shapes:

– **Square Well**

– **Ramped Well**

– **Asymmetric Well**

– **Double Square Well**

– **Parabolic Well**

– **Coulombic Well**

– **User Defined** (This item only appears if the **UserFlag** within the program itself; see section 2.5.2.)

● **Parameters**

– **Vary Well Parameters**
Allows you to change the parameters of the well. Choosing **[Ok]** or pressing <**Enter**> will accept whatever numbers appear in the boxes on the screen, unless they are outside the allowed range, in which case the program will give an error message and wait for the numbers to be re-entered. Choosing **[Cancel]** or pressing <**Esc**> will abort entry and return to the main menu, with the well parameters unchanged from what they were before. Choosing **[View]** displays the well with the numbers just entered, but does not accept them permanently.

– **Add a Perturbation**
This allows the user to add a small perturbation to the well already selected. Choices available are—

* **Constant**

* **Linear**

* **Quadratic**

* **Cubic**

* **Quartic**

* **User Defined** (The item appears only if the **UserFlag** is set; see section 2.5.2.)

The coefficient which has to be entered must be small enough that the perturbation is nowhere more than 10% of the maximum depth of the well. Choosing **[Ok]** or pressing <**Enter**> will accept the choice of perturbation type and whatever number appears in the coefficient box on the screen, unless the number is too large, in which case the program will give an error message and wait for the number to be re-entered. Choosing **[Cancel]** or pressing <**Esc**> will abort entry and return to the main menu, with the well unchanged from what it was before. Choosing **[View]** displays the well with the type and number just entered, but does not accept them permanently.

● **Method**
This part of the program allows you to solve Schrödinger equation for yourself. It is not strictly necessary if you are only interested in presenting physical results. It shows the methods by which those results were obtained, and will be useful if you make changes yourself to any of the program's parameters. It allows four different submenu choices:

– **Try Energy (With Mouse)**
This allows you to select an energy by clicking within the graph. The Schrödinger equation is then solved and the solution drawn on the screen. The value of the energy chosen also appears in a message box at the top of the screen, together with the number of nodes. *Note: This number might be one greater than the number of nodes you can actually see. That means there is another node off to the right somewhere.*

Three hot keys may be selected, with the mouse or by pressing—

* **Clear**   Redraw the screen without the previous solutions shown.

* **Zoom In/Out**   Increase/decrease the scale by a factor of two; useful when trying to home in on an eigenvalue.

* **Menu**   End entry and return to the main menu.

– **Try Energy (From Keyboard)**
This allows you to enter an energy directly. The Schrödinger equation is then solved and the solution drawn on the screen. *N.B.: The quantity you enter is actually E_B, and must be positive.* The value of the energy chosen also appears in a message box at the top of the screen, together with the number of nodes. *Note: This number might be one greater than the number of nodes you can actually see. That means there is another node off to the right somewhere. This is important when you enter a number which is only approximately equal to an eigenvalue. The shape of the function may look as you think it should, but since the "real" eigenvalue may be a little less than your choice, the program will find one more node than you expect (which may well be off-screen).*

Special results occur when the following choices are made.

* **[Clear]**   Redraw the screen without the previous solutions shown.

* **[Ok]**   Accept the number entered and show the solution, as just described.

* **[Cancel]**   End entry and return to the main menu.

– **Hunt for Zero**
This allows you to find accurate eigenvalues by having the program hunt in a binary search between two limits. When you choose this option a screen appears asking you to estimate a value for the binding energy which you think is too big (an upper bound). It integrates the equation and displays the solution. It then asks you for a number you think is too low (a lower bound). When you've supplied it, it again integrates and displays the solution. You may at any time select the number appearing in either box by choosing **[Ok]** or pressing <**Enter**>. If the asymptotes of these solutions are of opposite sign, it assumes that there is a zero between the two. It locates this zero by a binary chopping method, displaying the value of the energy and the shape of the "solution" of the equation for each value it tries.

When it has found the zero to sufficient accuracy, it clears the screen, waits for you to press any key, then draws the eigenfunction it has found and the probability distribution on the same plot, displays the value of the energy eigenvalue, and redraws the potential and marks on it the energy level.

During the first part of this procedure, however, you may not be able to supply appropriate upper and lower bounds. If you give it two numbers for which the asymptotes are of the same sign, it will beep and then check the number of nodes in the two solutions. If they differ (which means there is more than one eigenvalue in the range), it will find one of these. If they are the same, it will wait for you to input another pair of numbers. If you just cannot find a suitable pair of bounds — and it gets difficult when the energy levels are very close together — you can get out by choosing **[Cancel]** or pressing <**Escape**>. This puts you back to the main menu.

– **Examine Solution**
This allows you to read from the screen actual values of the wave function and the *x*-position, simply by pointing to the drawn curve with the mouse. The information appears in a message box at the top of the screen. Pressing or clicking on <**Esc**> returns you to the main menu. *Note: This menu item can only be selected immediately after having found a solution with* **Hunt for Zero** *or in a couple of other cases when there is unambiguously a wave function drawn on the screen capable of being examined.*

● **Spectrum**

– **Find Eigenvalues**
The program automatically searches for the eigenvalues. It does this by a recursive procedure, so they appear on the screen in no fixed order. As each is found, the energy level is drawn on the graph and a beep is sounded. This sound may be disabled if it is annoying (see **Sound** below).

– **See Wave Function**

This item displays the eigenfunctions previously found for the particular potential that has been chosen. *It cannot be selected if you have changed the potential well but have not yet found all the eigenvalues.* On first entry it asks you which of the eigenfunctions you want to see, thus:

Choose a level number ...... [Clear] [Ok] [Cancel]

Enter from the keyboard. If you try to give it a number greater than the number of eigenvalues, or if you enter an alphabetic character, it will beep at you and give an error message. When you have entered an acceptable number it will draw that eigenfunction, and display in the window the level number and the energy eigenvalue. It will also highlight on the potential plot the energy level in question. It will then wait for you to enter another level number. When you've seen enough eigenfunctions, choose **[Cancel]** or press $<$**Esc**$>$ and you will return to the main menu.

– **Examine Solution**

This allows you to read from the screen actual values of the last eigenfunction you plotted and the corresponding x-positions, simply by pointing to the drawn curve with the mouse. The information appears in a message box at the top of the screen. Pressing or clicking on $<$**Esc**$>$ returns you to the main menu. *Note: This menu item can only be selected immediately after having displayed one or more eigenfunctions with the previous two menu items or after* **Hunt for zero,** *when there also is unambiguously a wave function drawn on the screen capable of being examined.*

– **See Wfs and Probs**

This does the same as the previous menu item except that it will plot the probability density as well as the wave function each time.

– **Sound**

This toggles between the sound being turned on and turned off.

Part 2: Wave Function Properties

This part of the program can be used to calculate various overlap integrals involving the eigenfunctions you found in part 1, or linear combinations of them. Note especially that you must be sure that you have calculated the eigenfunctions before attempting to use these features (unless you have not changed the potential from its default form). In this part of the program the screen layout becomes that shown in Figure 2.2. The right − most four menu items have changed to the following.

● **Psi 1**

This allows you to choose the wave function which will be the first item in the overlap integrand. There are two possible choices:

– **Eigenstate, n = 1,2,3...**

Choose the eigenstate you want directly from the pull-down menu.

– **General State**

You are asked to supply coefficients for all of the bound state eigenfunctions

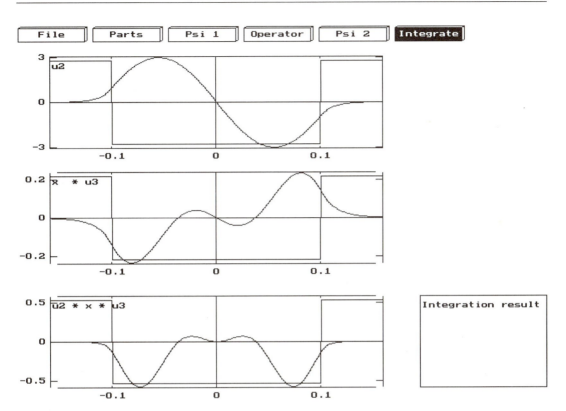

Figure 2.2: An illustration of the screen layout for part 2 of the program. The potential well being explored is the default square well. An overlap integrand is being constructed by multiplying the first excited state u_2 (top graph) by the operator x operating on the second excited state u_3 (second graph). The product is shown in the third graph. The area under this graph will be calculated and displayed in the fourth graph box.

for the well you are working with. None is allowed to be greater than 1. If you enter any number greater than this, it will give an error message and wait for you to re-enter a number in the right range. Choosing **[Cancel]** or pressing <**Esc**> will exit back to the main menu, leaving **Psi 1** unchanged.

- **Operator**

  This allows the operator in the middle of the overlap integrand to be specified. The choices are

  $$1 \quad x \quad d/dx \quad x^2 \quad d^2/dx^2 \quad V \quad E \quad x\,d/dx \quad d/dx\,x \quad user1 \quad user2\,.$$

  The last two items appear only if the **UserFlag** is set (see section 2.5.2). The operator currently in use is checked.

- **Psi 2**

  This allows you to choose the wave function which will be the last item in the overlap integrand. There are three possible choices:

  – **Eigenstate, n = 1, 2, 3 ...**

    Choose the eigenstate you want directly from the pull-down menu.

– **General state**

You are asked to supply coefficients for all of the bound state eigenfunctions for the well you are working with. None is allowed to be greater than 1.0000. If you enter any number greater than this it will give an error message and wait for you to re-enter a number in the right range. Choosing **[Cancel]** or pressing <**Esc**> will exit back to the main menu, leaving **Psi 2** unchanged.

– **Some Other Function**

This allows you to enter a completely unrelated function. A screen will ask you to enter the algebraic form of the function. Only a limited number of forms are accepted. The default value that appears is a "hat" function centered at $x = 0.075$ nm. Choosing **[Ok]** or pressing <**Enter**> accepts the algebraic form, provided it can be recognized by the CUPS parser. If it cannot an error message will appear and it will wait for further entry. Choosing **[Cancel]** or pressing <**Esc**> will exit back to the main menu, leaving **Psi 2** unchanged.

● **Integrate**

This will cause the overlap integration to be performed and the result to be shown in a separate box.

Part 3: Time Development

This part of the program is used to observe how linear combinations of eigenfunctions change with time. Note especially that you must be sure that you have calculated the eigenfunctions before attempting to use these features (unless you have not changed the potential from its default form). Upon entering this part of the program, you will find the right-most four menu items have been changed to the following layout, as shown in Figure 2.3.

● **Wave Func**

– **Choose Wave Function**

Allows you to enter coefficients (real and imaginary) to construct the general state at time $t = 0$.

– **Show Development**

Shows the time development by animation on the screen. The animation will continue until you click on one of the following hot keys.

* **F2 Run/Stop**

Start or stop the animation and wait.

* **F4 Forward/Reverse**

Set time step to the negative of what it was before.

* **F5 Slower**

Divide the time step by 2.

* **F6 Faster**

Multiply the time step by 2.

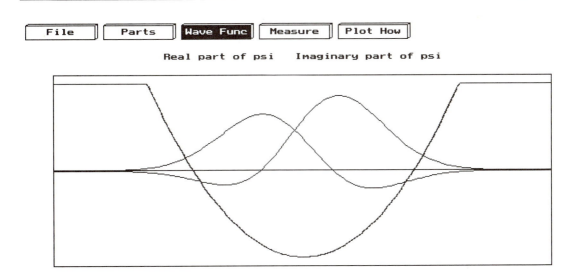

| File | Parts | Wave Func | Measure | Plot How |

Real part of psi   Imaginary part of psi

Time =   0.0180 (fs)

Figure 2.3: An illustration of the screen layout for part 3 of the program. The potential well being explored is a parabolic well. A general state at time $t = 0$ has been constructed by adding
together the ground state and the first excited state, each multiplied by a coefficient, 0.7071. The
real and imaginary parts of this general state are shown at time $t = 0.018$ fs.

* **F7 Restart**
  Set the time back to zero, leaving the time step unchanged.

* **F10 Menu**
  Stop animation and return to main menu.

– **Begin Over**
  Reset the time to zero and start the animation again.

● **Measure**

 – **Position**
   Calculate the expectation value of $\langle x \rangle$ at the current time.

 – **Momentum**
   Calculate the expectation value of $\langle p \rangle$ at the current time.

 – **User Defined**
   The item appears only if the **userFlag** is set (see section 2.5.2).

 – **Set Time**
   Set the time to any desired value.

● **Plot How**

 – **Amplitude and Phase**
   Plot the amplitude of the (complex) wave function, and use color to identify

the phase at any point. A color wheel is drawn at the side to help identification. This is the default display mode.

- **Real and Imaginary**
  Plot the real (red) and imaginary (green) parts of the wave function on the same graph.

2.5.2   Possible Modifications to the Program

As you become more used to working with this program, you may find the need to do calculations that the author did not think of. The program is designed to make it easy for you to go inside and add extra pieces of code of your own. For example, it is very straightforward to add a new potential well shape.

You may, of course, change any part of the code you choose, but there are a number of "user defined" procedures already written which you may use as templates. These are gathered together in the part of the program headed by the comment

********* USER DEFINED PROCEDURES ********

In that part of the code there is a Boolean variable **userFlag**. When you first get the program it is set to be *false*, but if you set it to *true*, then at several points in the program an extra menu item will appear, called something like **User Defined**. By selecting that item you may access parts of the code which you yourself write.

There are four places where this occurs. In each case you are advised to write the specified procedure you want by using the template that is there already, although you do not have to. You may, of course, use the existing template procedures if they do what you want. *Note, however, that they will flash a warning message telling you that this is only a default, and you will have to write your own code. You will need to comment out this* **Announce** *statement if you don't want it to keep interrupting you.*

1. **Part 1. Potential | User Defined**
   If you write a procedure called **CalculateUserDVector**, you can incorporate your own well. The template provided is for a triangular well.

2. **Part 1. Parameters | Add a Perturbation | User Defined**
   If you write a procedure called **SetUserPert**, you can add your own perturbation to the potential.

3. **Part 2. Operator | User1/User2**
   If you write a procedure called **ConstructUserKet**, you can incorporate any two operations you want to be carried out on the function **Psi 2**. The templates provided calculate the operation of $(a.x \pm b.d/dx)$ on **Psi 2**. These are ladder operators for the parabolic well eigenfunctions (and the two constants a and b are chosen to fit that purpose).

4. **Part 3. Measure | User Defined**
   If you write a procedure called **MeasureUserQuant** you can get the

program to calculate any expectation value (or to perform any other calculation you like) using the (complex) wave function at the current time **Psi 2**. The template provided calculates $\langle x^2 \rangle$.

Note that, if you *really* want to customize the program to your own needs, you may also change the name "User Defined" which appears in the menu items to something of your own choosing. Just follow the templates in the procedure **Set-UserNames**.

Here are suggestions to show what you might do.

- Construct a double harmonic oscillator.
  Measure the energy difference of pairs of states. Compare with theoretical values (see the theoretical description in Merzbacher,[1] section 5.6).

- Add a ladder operator for the harmonic oscillator well.
  Demonstrate that its effect on eigenfunction shapes is what it should be. Find the multiplying constant it produces and compare with the theoretical value. (This code has been written already, and will be called into play if you set the **userFlag** equal to *true*. It is, however, not incorporated into the main program because it is a very specialized application.)

References

1. Merzbacher, E. *Quantum Mechanics*, 2nd ed. New York: John Wiley & Sons, 1970.

2. Friar, J. L. A Note on the Roundoff Error in the Numerov Algorithm. Journal of Computational Physics **28:**426, 1978.

3. French, A. P., Taylor, E. F. *Introduction to Quantum Mechanics*. London: Thomas Nelson & Sons, 1978.

4. Haken, H. *Light: Volume 1: Waves, Photons, Atoms*. Amsterdam: North-Holland Publishing Co., 1981.

5. Press, W. H., Flannery, B. P., Teukolsky, S. A., Vetterling, W. T. *Numerical Recipes: The Art of Scientific Computing in Pascal*. New York: Cambridge University Press, 1988.

3

Stationary Scattering States in One Dimension

John R. Hiller

3.1 Introduction

One of the most fundamental types of experiments on the quantum level is the scattering experiment. It is used to explore the structure of atoms and nuclei, and even of nucleons themselves. In its basic form such an experiment consists of a beam of particles directed at a target and one or more detectors placed to observe the direction of scattered particles. In more complicated experiments, the result of the beam-target interaction may be creation of several particles and target fragments. Observations of the results provide clues to the structure of the target, the nature of its interaction with the beam particles, and the structure of the beam particle, if any.

Here we will consider a simple situation where the beam particle has no structure and can move only along a straight line. To be specific, we assume the particle is an electron. Also, the target is represented by a potential energy function with no imaginary part. During the scattering process, total energy is then conserved. The scale of the potential is taken to be on the order of electron volts. Because motion occurs only along a line, one can only ask whether the electron is reflected backward from the target or transmitted forward. It is the nature of quantum mechanics that these outcomes are predicted as probabilities only. An example of a situation where these considerations can be applied is the tunneling of conduction electrons through insulating barriers at junctions.

The analysis is done in terms of states with well-defined energies. Of course, this is an idealization, since one cannot generate a beam of particles where each particle has precisely the same energy. However, if a prediction can be made for the transmission and reflection probabilities as functions of energy, then any uncertainties in the beam particle energy can be included as a last step.

The remaining sections describe the analysis and the program **Scattr1D** that can be used to compute results for a wide range of situations. While running the program, the user can control the potential energy, and obtain plots of the transmission and reflection probabilities versus energy and plots of the wave function for the state at a particular energy.

The scattering process can also be analyzed in terms of wave packets. The packet is constructed with a given average momentum and launched toward the target. Its behavior is then dictated by the time-dependent Schrödinger equation. Such a process can be observed by running the program **QMTime**, which is discussed in chapter 4.

3.2 *Scattering in One Dimension*

3.2.1 Stationary States

The state of a particle in a potential $V(x)$ is described by a wave function $\Psi(x, t)$, which must satisfy the time-dependent Schrödinger equation:

$$-\frac{\hbar^2}{2m}\frac{\partial^2}{\partial x^2}\Psi + V(x)\Psi = i\hbar\frac{\partial}{\partial t}\Psi. \tag{3.1}$$

A stationary state of this particle is one where the probability density $|\Psi(x, t)|^2$ is independent of time. This does not necessarily mean that the wave function $\Psi(x, t)$ is also independent of time. By separation of variables, one obtains

$$\Psi(x, t) = \psi(x)e^{-iEt/\hbar}, \tag{3.2}$$

where ψ must obey the time-independent Schrödinger equation,

$$-\frac{\hbar^2}{2m}\frac{d^2\psi}{dx^2} + V(x)\psi = E\psi, \tag{3.3}$$

and E is the energy. For the remainder of the chapter, we will refer to the time-independent Schrödinger equation as *the* Schrödinger equation, unless confusion might arise.

The type of stationary states with which this chapter is concerned are those for which E is above the value of the potential energy $V(x)$ at either $x = \infty$ or $x = -\infty$ or both. These are states that can be used to describe scattering processes where particles are incident from some direction with finite kinetic energy.[1-7] Stationary states that are bound are discussed in chapter 2. This distinction is made more clear in the discussion of boundary conditions given below.

3.2.2 Properties of Scattering States

Regions of Constant Potential Energy

The Schrödinger equation (Eq. 3.3) is a second-order differential equation, and, therefore, one can find two linearly independent solutions. Any general solution can then be constructed as a linear combination of these. In a region where the potential $V(x)$ is constant, pairs of solutions can be determined analytically. If E is greater than V, we have

$$\frac{d^2\psi}{dx^2} = -k^2\psi,$$ (3.4)

with $k^2 = 2m(E - V)/\hbar^2 > 0$. Of course, $\hbar k$ is the classical momentum. The solutions may be chosen to be $\sin kx$ and $\cos kx$, or e^{ikx} and e^{-ikx}. If E is less than V, then

$$\frac{d^2\psi}{dx^2} = \kappa^2\psi,$$ (3.5)

with $\kappa^2 = 2m(V - E)/\hbar^2 > 0$, and two solutions are $e^{\kappa x}$ and $e^{-\kappa x}$. In the special case of equality between E and V, the equation is simply

$$\frac{d^2\psi}{dx^2} = 0,$$ (3.6)

and the solutions are x and a constant. To summarize, the general solution to the Schrödinger equation in a region of constant V may be written as

$$\psi = \begin{cases} \begin{rcases} \begin{array}{c} A \sin kx + B \cos kx \\ \text{or} \\ Ae^{ikx} + Be^{-ikx} \end{array} \end{rcases} , E > V \\ A + Bx \qquad\qquad , E = V \\ Ae^{-\kappa x} + Be^{\kappa x} \qquad , E < V, \end{cases}$$ (3.7)

with A and B constants that are arbitrary and possibly complex.

Probability Current Density

In order to interpret these solutions, we introduce the notion of a probability current density $j(x, t)$, which quantifies the flow of probability density from one region to another as an amount per unit time. Such a flow of probability is what one would expect to be associated with the motion of a particle.

Consider an interval $[x, x + \Delta x]$. The probability of a particle being in this interval is approximated by the product of the probability density $\rho(x, t) = |\Psi(x, t)|^2$ and the interval length Δx. There are two ways for this probability to change with time: through explicit time dependence (which is absent for a stationary state) and through the flow of probability across the ends of the interval. The change due to explicit time dependence is $\rho(x, t + \Delta t)\Delta x - \rho(x, t)\Delta x$. The increase due to flow from the left is $j(x, t)\Delta t$ and the decrease due to flow to the right is $j(x + \Delta x, t)\Delta t$. Note that positive j means flow to the right.

Conservation of probability requires that a local increase or decrease be due to a net inflow or outflow; therefore, we have

$$\rho(x, t + \Delta t)\Delta x - \rho(x, t)\Delta x = j(x, t)\Delta t - j(x + \Delta x, t)\Delta t.$$ (3.8)

In the limit where both Δt and Δx go to zero, this yields the standard continuity equation

$$\frac{\partial \rho}{\partial t} + \frac{\partial j}{\partial x} = 0 \,. \tag{3.9}$$

This can be used to relate j to the wave function Ψ.

To make the desired connection, we return to the explicit time dependence and express $\partial \rho / \partial t$ directly in terms of the wave function

$$\frac{\partial \rho}{\partial t} = \left(\frac{\partial \Psi}{\partial t} \right)^* \Psi + \Psi* \left(\frac{\partial \Psi}{\partial t} \right) \,. \tag{3.10}$$

Use of the time-dependent Schrödinger equation (Eq. 3.1) yields

$$\frac{\partial \rho}{\partial t} = \left(-\frac{i}{\hbar} \left[-\frac{\hbar^2}{2m} \frac{\partial^2 \Psi}{\partial x^2} + V(x)\Psi \right] \right)^* \Psi + \Psi* \left(-\frac{i}{\hbar} \left[-\frac{\hbar^2}{2m} \frac{\partial^2 \Psi}{\partial x^2} + V(x)\Psi \right] \right), \tag{3.11}$$

which, since V is real, can be reduced to

$$\frac{\partial \rho}{\partial t} = -i\frac{\hbar}{2m} \left(\frac{\partial^2 \Psi*}{\partial x^2} \Psi - \Psi* \frac{\partial^2 \Psi}{\partial x^2} \right) \,. \tag{3.12}$$

To extract an expression for j, one need only write the right-hand side of Eq. 3.12 as a total derivative and compare the result with Eq. 3.9. This yields

$$j = \frac{i\hbar}{2m} \left(\frac{\partial \Psi*}{\partial x} \Psi - \Psi* \frac{\partial \Psi}{\partial x} \right) \,. \tag{3.13}$$

In the case of the stationary states for which $\Psi = \psi \exp(-iEt/\hbar)$, this reduces to

$$j = \frac{i\hbar}{2m} \left(\frac{\partial \psi*}{\partial x} \psi - \psi* \frac{\partial \psi}{\partial x} \right) \,. \tag{3.14}$$

The more compact form, $j = \mathcal{R}e[\psi*(p/m)\psi]$, shows the close correspondence to the classical form $j_{cl} = \rho_{ch} v$, where the charge density ρ_{ch} is the analog of $\psi*\psi$, and $v = p/m$ is the velocity.

One then immediately sees that a non-zero current density can arise only when ψ is complex. The plane-wave solutions Ae^{ikx} and Ae^{-ikx} in the constant-potential case are such wave functions. The current is computed to be

$$j = \pm |A|^2 \frac{\hbar k}{m} \quad \text{for} \quad \psi = Ae^{\pm ikx}. \tag{3.15}$$

Thus, for positive k, e^{ikx} corresponds to current moving right and e^{-ikx} to current moving left. For negative k, the directions reverse. In this simple case, the magnitude of the current is given by the probability density $\rho = |A|^2$ times the classical velocity $\hbar k/m$.

Boundary Conditions

We can now readily define boundary conditions appropriate to a stationary scattering state. Beyond some finite range in x, the potential $V(x)$ is assumed constant. Particles with energies greater than this constant level may then propagate to and from infinity with finite momentum. Without loss of generality, the zero in the energy scale may be shifted to make the potential zero in the direction from which the scattered beam is incident. The beam energy may then take on any positive value. Again, without loss of generality, we can choose the beam direction to be to the right.

Given these assumptions, an incident particle of energy E will be associated with a plane-wave solution e^{ikx}, with $k = \sqrt{2mE}/\hbar$ far to the left. A reflected particle will be associated with e^{-ikx}. Energy conservation implies that this k is the same k as for the incident particle. Thus, to the left of the target region, the wave function is a linear combination of e^{ikx} and e^{-ikx}.

A transmitted wave will take one of three forms. If E is greater than $V(\infty)$, it will be proportional to $e^{ik'x}$, where $k' = \sqrt{2m[E - V(\infty)]}/\hbar$. If E is less than or equal to $V(\infty)$, transmission does not take place, and the wave dies out. The solutions that grow with x must be excluded. If E is strictly less than $V(\infty)$, we use $e^{-\kappa x}$, with $\kappa = \sqrt{2m[V(\infty) - E]}/\hbar$; and if E is equal to $V(\infty)$, we use a constant. In either of these last two cases, the transmitted current density is zero.

In the program **Scattr1D**, these choices of boundary conditions are laid out as follows. The potential is allowed to vary only in an interval $[x_{min}, x_{max}]$. Outside this interval, V is zero to the left and equal to a constant V_∞ to the right. These constraints on V are illustrated in Figure 3.1. The boundary conditions on the wave function are then

$$\psi = \begin{cases} Ae^{ikx} + Be^{-ikx} & , x < x_{min} \\ \left. \begin{array}{l} Ce^{ik'x} \ , E \geq V_\infty \\ Ce^{-\kappa x}, E < V_\infty \end{array} \right\} & , x > x_{max}. \end{cases} \tag{3.16}$$

Of the coefficients A, B, and C, one can be arbitrarily chosen, because the Schrödinger equation is linear and homogeneous. The other two are determined in the process of finding the solution.

The full solution is computed by solving the Schrödinger equation in the interior and matching the boundary conditions at x_{min} and x_{max}. If the potential reaches a height that is greater than the energy E, transmission can still take place as a tunneling process, provided $E > V_\infty$. In the region where the energy is below the level of the potential, the wave function is exponentially damped. Over any finite range, the probability density does drop but it never reaches zero.

The boundary conditions for scattering states differ markedly from those for bound states. In the latter case, the particle does not escape; therefore, the wave function must tend to zero in both directions.

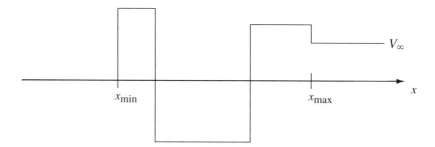

Figure 3.1: An example of a potential that satisfies the constraints required by the program.

Transmission and Reflection Probabilities

For an experiment conducted during a period of time τ, the fraction T of particles transmitted is given by the ratio of integrated currents:

$$T = \frac{\left| \int_0^\tau j_{\text{trans}}\, dt \right|}{\left| \int_0^\tau j_{\text{inc}}\, dt \right|}. \tag{3.17}$$

Here j_{inc} is the incident current density and j_{trans} the transmitted current density. The fraction R of particles reflected is similarly given by

$$R = \frac{\left| \int_0^\tau j_{\text{refl}}\, dt \right|}{\left| \int_0^\tau j_{\text{inc}}\, dt \right|}, \tag{3.18}$$

with j_{refl} the reflected current density. If we assume that the experiment is arranged to have a constant incident current density (a constant particle flux), all the currents are constant and the time intervals disappear from the ratios. From the boundary conditions given in Eq. 3.16, we obtain

$$j_{\text{trans}} = \begin{cases} \frac{\hbar k'}{m}|C|^2 & , E > V_\infty \\ 0 & , E \le V_\infty; \end{cases} \tag{3.19}$$

and from the incident and reflected pieces we find

$$j_{\text{inc}} = \frac{\hbar k}{m}|A|^2, \quad j_{\text{refl}} = \frac{\hbar k}{m}|B|^2. \tag{3.20}$$

These lead directly to the result

$$T = \frac{k'}{k}\left|\frac{C}{A}\right|^2, \quad R = \left|\frac{B}{A}\right|^2. \tag{3.21}$$

Conservation of probability requires that $R + T = 1$. This condition must be satisfied by the solution for any real potential.

3.2.3 Scattering Solutions of the Time-Independent Equation

There are some potentials for which an analytic solution can be obtained. One class of such potentials consists of those that are piecewise constant. To understand better the numerical solutions obtained by **Scattr1D**, it is instructive to study the process of analytic solution. Two examples are the step potential

$$V_{\text{step}} = \begin{cases} 0 & , x < 0 \\ V_0 & , x > 0 \end{cases} \tag{3.22}$$

and the square barrier

$$V_{\text{sqr}} = \begin{cases} 0 & , x < -a \\ V_0 & , -a < x < a \\ 0 & , x > a, \end{cases} \tag{3.23}$$

where V_0 is a positive constant.

Step Potential

Consider first the step potential. We already have the general form of the solution on either side of the step at $x = 0$, and all that remains is to determine the coefficients that appear. For definiteness, let us assume that the energy is greater than V_0. Then we have

$$\psi = \begin{cases} Ae^{ikx} + Be^{-ikx} & , x < 0 \\ Ce^{ik'x} & , x > 0, \end{cases} \tag{3.24}$$

with $k = \sqrt{2mE}/\hbar$ and $k' = \sqrt{2m[E - V_0]}/\hbar$. Two (complex) conditions are now needed to fix the ratios of coefficients. These come from the continuity of ψ and $\psi' \equiv (d\psi/dx)$ across the step.

Once we assume that the Schrödinger equation holds at $x = 0$, ψ must be continuous there. The continuity of ψ' is then also determined by the Schrödinger equation. One need only integrate the equation over the interval $[-\epsilon, \epsilon]$ to find

$$\psi'(x = \epsilon) - \psi'(x = -\epsilon) = -\frac{2m}{\hbar^2} \int_{-\epsilon}^{\epsilon} dx[E - V(x)]\psi(x) \tag{3.25}$$

and take the limit $\epsilon \to 0$. Since V is well behaved, the right-hand side has a limit of zero.

The continuity conditions reduce to the following algebraic conditions:

$$A + B = C, \quad ik(A - B) = ik'C. \tag{3.26}$$

They may be combined to obtain

$$\frac{B}{A} = \frac{k - k'}{k + k'}, \quad \frac{C}{A} = \frac{2k}{k + k'}. \tag{3.27}$$

The transmission and reflection probabilities are then

$$T = \frac{4kk'}{(k + k')^2}, \quad R = \left(\frac{k - k'}{k + k'}\right)^2. \tag{3.28}$$

Notice that they do satisfy the condition $R + T = 1$.

Square Barrier

Obtaining the solution for the square barrier is nearly as simple. This time we consider the case where E is less than V_0, but still positive, of course. The wave function is

$$\psi = \begin{cases} A_0 e^{ikx} + B_0 e^{-ikx} & , x < -a \\ A_1 e^{-\kappa x} + B_1 e^{\kappa x} & , -a < x < a \\ A_2 e^{ikx} & , x > a, \end{cases} \tag{3.29}$$

where $k = \sqrt{2mE}/\hbar$ and $\kappa = \sqrt{2m[V_0 - E]}/\hbar$. Continuity conditions are applied at both $-a$ and a; they yield

$$A_0 e^{-ika} + B_0 e^{ika} = A_1 e^{\kappa a} + B_1 e^{-\kappa a}$$

$$ik(A_0 e^{-ika} - B_0 e^{ika}) = -\kappa(A_1 e^{\kappa a} - B_1 e^{-\kappa a})$$

$$A_1 e^{-\kappa a} + B_1 e^{\kappa a} = A_2 e^{ika} \tag{3.30}$$

$$-\kappa(A_1 e^{-\kappa a} - B_1 e^{\kappa a}) = ik A_2 e^{ika}.$$

One approach to solving these algebraic equations is to work from right to left in x, by first finding A_1 and B_1 in terms of A_2:

$$A_1 = \left(1 - i\frac{k}{\kappa}\right) e^{ika} e^{\kappa a} \frac{A_2}{2},$$

$$B_1 = \left(1 + i\frac{k}{\kappa}\right) e^{ika} e^{-\kappa a} \frac{A_2}{2}. \tag{3.31}$$

Substitution into the first two equations in Eq. 3.30 yields

$$A_0 e^{-ika} + B_0 e^{ika} = \frac{A_2}{2} e^{ika}\left[\left(1 + i\frac{k}{\kappa}\right) e^{-2\kappa a} + \left(1 - i\frac{k}{\kappa}\right) e^{2\kappa a}\right],$$

$$ik(A_0 e^{-ika} - B_0 e^{ika}) = \frac{A_2}{2} \kappa e^{ika}\left[\left(1 + i\frac{k}{\kappa}\right) e^{-2\kappa a} - \left(1 - i\frac{k}{\kappa}\right) e^{2\kappa a}\right]. \tag{3.32}$$

Finally, we obtain

$$\frac{A_2}{A_0} = 4ik\kappa e^{-2ika}[(\kappa + ik)^2 e^{-2\kappa a} - (\kappa - ik)^2 e^{2\kappa a}]^{-1} \tag{3.33}$$

and

$$\frac{B_0}{A_0} = (k^2 + \kappa^2) e^{-2ika}(e^{2\kappa a} - e^{-2\kappa a})[(\kappa + ik)^2 e^{-2\kappa a} - (\kappa - ik)^2 e^{2\kappa a}]^{-1}. \tag{3.34}$$

We introduce hyperbolic sines and cosines to simplify these expressions somewhat:

$$\frac{A_2}{A_0} = 2ik\kappa e^{-2ika}[(k^2 - \kappa^2) \sinh 2\kappa a + 2ik\kappa \cosh 2\kappa a]^{-1}, \tag{3.35}$$

$$\frac{B_0}{A_0} = (k^2 + \kappa^2) e^{-2ika} \sinh 2\kappa a [(k^2 - \kappa^2) \sinh 2\kappa a + 2ik\kappa \cosh 2\kappa a]^{-1}. \tag{3.36}$$

The transmission and reflection probabilities are then given by

$$T = \frac{4k^2 \kappa^2}{(k^2 - \kappa^2) \sinh^2 2\kappa a + 4k^2 \kappa^2 \cosh^2 2\kappa a}, \tag{3.37}$$

$$R = \frac{(k^2 + \kappa^2)^2 \sinh^2 2\kappa a}{(k^2 - \kappa^2)^2 \sinh^2 2\kappa a + 4k^2 \kappa^2 \cosh^2 2\kappa a}. \tag{3.38}$$

Of course, classically the transmission probability would be zero. Here zero is obtained only in the limit of $V_0 \gg E$ and $2mV_0 a^2/\hbar^2 \gg 1$; these two conditions are necessary for κa to be large, and they correspond to having a barrier which is both tall and wide.

Potentials of Arbitrary Shape

Even when the potential is not piecewise constant, there are still two linearly independent solutions to the Schrödinger equation. In certain cases they may be found among functions that have already been studied and their properties cataloged. Let ψ_a and ψ_b be two linearly independent solutions, and form the linear combination $A_1\psi_a + B_1\psi_b$ with unknown coefficients. This combination must satisfy the continuity conditions at $x_{\min}$ and $x_{\max}$ that allow it to match the boundary conditions given in Eq. 3.16. One again obtains algebraic equations that determine all of the coefficients, up to a multiplicative complex constant.

3.3   Computational Approach

The program **Scattr1D** can solve the scattering-state problem in one of three ways. The first applies a piecewise-constant approximation to the potential, and uses general formulas for piecewise-constant potentials. If the original potential is piecewise constant, this method may be used in a special form that makes no approximation for the potential and solves the problem exactly. The second method uses a direct numerical integration of the Schrödinger equation. The third also uses numerical integration, but for the logarithm of the wave function. This algorithm suffers from an instability when E is near V_∞; however, it works well in the domain of small transmission probabilities where the direct numerical integration method does not.

In each of the approximate methods, the interval $[x_{\min}, x_{\max}]$ is divided into a number of equal-length subintervals called steps. The size of the step has a direct effect on the accuracy of the approximation. In particular, the wave function oscillates in the classically allowed regions and is poorly approximated when the step size is not significantly smaller than the period of the oscillation. This means that study of larger energies requires smaller step sizes. The user of the program is responsible for selecting an appropriate number of steps, which can be done using the dialog box obtained when the **Parameters** menu item is chosen.

3.3.1   Scaled Variables

To keep the results within a reasonable range of values, the simulation is written to work with the Schrödinger equation in a dimensionless form. Energies and lengths are measured relative to scales V_0 and L_0, respectively. Input and output are handled in terms of quantities with dimensions, which are translated using these scale factors. The scaled quantities are

$$\bar{x} = x/L_0, \quad \bar{E} = E/V_0, \quad \bar{V}(\bar{x}) = V(\bar{x}L_0)/V_0, \tag{3.39}$$

and the Schrödinger equation becomes

$$-\frac{d^2\psi}{d\bar{x}^2} + \zeta\bar{V}(\bar{x})\psi = \zeta\bar{E}\,\psi, \tag{3.40}$$

where

$$\zeta \equiv 2mV_0L_0^2/\hbar^2 \tag{3.41}$$

is a dimensionless parameter. The mass of the particle enters the calculation only here. The scaled wave number is $\bar{k} \equiv kL_0 = \sqrt{2mL_0^2 E}/\hbar = \sqrt{\zeta \bar{E}}$.

The program computes results for electrons interacting with atomic-scale potentials. Therefore, the energy unit has been chosen to be the electron volt (eV) and the length scale to be $L_0 = 1$ nm.

3.3.2 Piecewise-Constant Potentials

The general case of a piecewise-constant potential is in principle no more difficult to solve than the two previous examples of the step and square barrier. We define the shape of the potential in the following way:

$$V = \begin{cases} 0 & , x < x_{\min} \equiv x_0 \\ V_0 & , x_{\min} < x < x_1 \\ V_1 & , x_1 < x < x_2 \\ \cdot \\ \cdot \\ \cdot \\ V_N & , x_N < x < x_{\max} \\ V_\infty & , x_{N+1} \equiv x_{\max} < x. \end{cases} \tag{3.42}$$

In each interval $[x_n, x_{n+1}]$, the solution takes one of the three standard forms:

$$\psi = \begin{cases} A_{n+1} e^{ikx} + B_{n+1} e^{-ikx} & , E > V_n \\ A_{n+1} + B_{n+1} x & , E = V_n \\ A_{n+1} e^{-\kappa x} + B_{n+1} e^{\kappa x} & , E < V_n. \end{cases} \tag{3.43}$$

At each endpoint x_n, continuity conditions are applied to ψ and ψ'. These conditions relate the coefficients on one side of x_n to the coefficients on the other. Such a relationship is conveniently expressed in matrix form

$$M_n(x_n) \begin{pmatrix} A_n \\ B_n \end{pmatrix} = M_{n+1}(x_n) \begin{pmatrix} A_{n+1} \\ B_{n+1} \end{pmatrix}, \tag{3.44}$$

where $M_n(x)$ is a complex matrix, the form of which depends on E and V_n. For $E > V_n$, we have

$$M_n(x) = \begin{pmatrix} e^{ik_n x} & e^{-ik_n x} \\ ik_n e^{ik_n x} & -ik_n e^{-ik_n x} \end{pmatrix}, \tag{3.45}$$

with $k_n = \sqrt{2m[E - V_n]}/\hbar$; for $E = V_n$,

$$M_n(x) = \begin{pmatrix} 1 & x \\ 0 & 1 \end{pmatrix}; \tag{3.46}$$

for $E < V_n$,

$$M_n(x) = \begin{pmatrix} e^{-\kappa_n x} & e^{\kappa_n x} \\ -\kappa_n e^{-\kappa_n x} & \kappa_n e^{\kappa_n x} \end{pmatrix}, \tag{3.47}$$

with $\kappa_n = \sqrt{2m[V_n - E]}/\hbar$.

We can write down a formal solution for A_0 and B_0 in terms of A_{N+1} and $V_{N+1} \equiv V_\infty$

$$\begin{pmatrix} A_0 \\ B_0 \end{pmatrix} = M_0^{-1}(x_0)M_1(x_0)M_1^{-1}(x_1)\cdots M_N(x_N)M_{N+1}^{-1}(x_N)M_{N+1}(x_{N+1})\begin{pmatrix} A_{N+1} \\ 0 \end{pmatrix}. \quad (3.48)$$

The ratios A_0/A_{N+1} and B_0/A_{N+1} are then determined by the products of matrices, and one can extract the transmission and reflection probabilities.

As shown by Kalotas and Lee,[8] the combination $K_n \equiv M_n(x_{n-1})M_n^{-1}(x_n)$, the transfer matrix, can be considerably simplified. The particular form again depends on the energy. For $E > V_n$, it is

$$K_n = \begin{pmatrix} \cos k_n(x_n - x_{n-1}) & -k_n^{-1}\sin kn(x_n - x_{n-1}) \\ k_n \sin k_n(x_n - x_{n-1}) & \cos k_n(x_n - x_{n-1}) \end{pmatrix}; \quad (3.49)$$

for $E = V_n$,

$$K_n = \begin{pmatrix} 1 & -1 \\ 0 & 1 \end{pmatrix}; \quad (3.50)$$

for $E < V_n$,

$$K_n = \begin{pmatrix} \cosh \kappa_n(x_n - x_{n-1}) & -\kappa_n^{-1}\sinh \kappa_n(x_n - x_{n-1}) \\ -\kappa_n \sinh \kappa_n(x_n - x_{n-1}) & \cosh \kappa_n(x_n - x_{n-1}) \end{pmatrix}. \quad (3.51)$$

In practice some care must be taken when E is near V_n; the denominators of the ratios $\sin k_n(x_n - x_{n-1})/k_n$ and $\sinh \kappa_n(x_n - x_{n-1})/\kappa_n$ become quite small, and the ratios are best computed from Taylor series.[8] Also, to accurately compute the transmission probability when it is small, the logarithm of the exponential factors $e^{-\kappa(x_{n+1}-x_n)}$ should be accumulated separately.

3.3.3   Piecewise-Constant Approximation

The general approach to piecewise-constant potentials can serve as a means to approximate the solution for any well-behaved potential.[8,9] One need only approximate the given potential by a piecewise-constant potential. For example, choose a grid of equally spaced points x_n between $x_{\min}$ and $x_{\max}$, and to match Eq. 3.42 define $V_n = V([x_n + x_{n+1}]/2)$. The program uses just this type of grid in procedure **SolveSchrodinger1**; however, one should realize that the choice of equal-length intervals is not necessary. A more efficient approximation might involve intervals carefully tailored to be long where V varies slowly and short where it varies rapidly.

The procedure **SolveSchrodinger1** does not explicitly use the transfer matrix K_n because the wave function is needed in addition to the transmission and reflection probabilities. The code works directly with the matrix M_n and its inverse. (See Eq. 3.48.) The coefficients are computed from right to left and the value of the wave function computed at the chosen steps.

3.3.4   Direct Integration

An alternative means of approximation is to integrate numerically the Schrödinger equation.[10] This is done in procedure **SolveSchrodinger2**. The transmitted wave, with its coefficient temporarily fixed at some arbitrary value, serves as the starting point for the integration, which extends from $x_{\max}$ to $x_{\min}$. The calculation can be done in such a way as to provide estimates for both ψ and ψ'. This is accomplished by working with the following system of first-order equations:

$$\frac{d\psi}{d\bar{x}} = \psi',$$

$$\frac{d}{d\bar{x}}\psi' = -\zeta[\bar{E} - \bar{V}(\bar{x})]\psi, \qquad (3.52)$$

which is equivalent to the Schrödinger equation. The starting values are

$$\psi(\bar{x}_{\max}) = e^{-30},$$

$$\psi'(\bar{x}_{\max}) = \begin{cases} ie^{-30}\sqrt{\zeta[\bar{E} - \bar{V}_\infty]} & ,\bar{E} > \bar{V}_\infty \\ 0 & ,\bar{E} = \bar{V}_\infty \\ -e^{-30}\sqrt{\zeta[\bar{V}_\infty - \bar{E}]} & ,\bar{E} < \bar{V}_\infty. \end{cases}$$

The choice of e^{-30} is somewhat arbitrary and is controlled by a constant **fInit** set to -30; this value gives the integration routine a large dynamic range with which to work. The form of $\psi'(\bar{x}_{\max})$ follows from differentiation of Eq. 3.16.

At $\bar{x}_{\min}$, the continuity conditions provide the means to extract A and B. The equations are

$$Ae^{i\bar{k}\bar{x}_{\min}} + Be^{-i\bar{k}\bar{x}_{\min}} = \psi(\bar{x}_{\min}),$$

$$i\bar{k}(Ae^{i\bar{k}\bar{x}_{\min}} - Be^{-i\bar{k}\bar{x}_{\min}}) = \psi'(\bar{x}_{\min}). \qquad (3.53)$$

The coefficient ratios and the transmission and reflection probabilities can then be computed.

3.3.5 Integration of the Logarithm

One situation where the use of scaled quantities does not keep the calculation within a reasonable range is one where the transmission probability is very small. This corresponds to a wave function that increases rapidly during integration from $x_{\max}$ to $x_{\min}$. The dynamic range necessary is then quite large.

To reduce the numerical range necessary, a natural approach is to compute the logarithm of the wave function, instead of the wave function itself. Procedure **SolveSchrodinger3** carries this out. We write the wave function as $e^{f(\bar{x})+ig(\bar{x})}$, with f and g real, and from the real and imaginary parts of the Schrödinger equation obtain the following coupled set of differential equations:

$$f'' + (f')^2 - (g')^2 = \zeta[\bar{V}(\bar{x}) - \bar{E}],$$

$$g'' + 2f'g' = 0. \qquad (3.54)$$

The second equation can be integrated once, to yield

$$g' = De^{-2f}, \qquad (3.55)$$

where D is an integration constant to be fixed by boundary conditions. The set of equations can then be written as three first-order equations:

$$\frac{df}{d\overline{x}} = F,$$

$$\frac{dF}{d\overline{x}} = -F^2 + G^2 + \zeta[\overline{V}(\overline{x}) - \overline{E}], \tag{3.56}$$

$$\frac{dg}{d\overline{x}} = G, \tag{3.57}$$

with G defined to be De^{-2f}.

The integration of this coupled set begins at $\overline{x}_{\max}$, where the values of the functions are fixed at

$$f(\overline{x}_{\max}) = -30,$$

$$F(\overline{x}_{\max}) = \begin{cases} 0 & , \overline{E} \geq \overline{V}_\infty \\ -\sqrt{\zeta[\overline{V}_\infty - \overline{E}]} & , \overline{E} < \overline{V}_\infty \end{cases} \tag{3.58}$$

$$G(\overline{x}_{\max}) = \begin{cases} -\sqrt{\zeta[\overline{V}_\infty - \overline{E}]} & , \overline{E} > \overline{V}_\infty \\ 0 & , \overline{E} \leq \overline{V}_\infty. \end{cases}$$

The constant D is given by

$$D = \begin{cases} e^{-60}\sqrt{\zeta[\overline{V}_\infty - \overline{E}]} & , \overline{E} > \overline{V}_\infty \\ 0 & , \overline{E} \leq \overline{V}_\infty. \end{cases} \tag{3.59}$$

These conditions follow from Eq. 3.53 and a choice of zero for $g(\overline{x}_{\max})$, to which multiples of 2π could be added with no practical effect. At $\overline{x}_{\min}$, the integration ceases and the coefficients of the incident and reflected waves are computed. The logarithm of the transmission probability is computed directly from f, F, and G.

This particular algorithm fails when E is very near V_∞. The integration of the nonlinear coupled set Eq. 3.56 becomes unstable. Its use should be limited to situations where T is very small and an accurate result for log T is desired.

3.4   Exercises

Many different calculations can be done with the program **Scattr1D**. A few are listed here as suggested exercises. Some require modification of the program. For those exercises that require addition of a new potential, the necessary steps are discussed in comments at the beginning of the program unit **Sc1DPotl**. Before trying the exercises, review the section on running the program.

3.1   **Program Check**

Compare results obtained with the program with those obtained analytically for a step, square barrier, and square well. For barriers, consider energies both above and below the top of the barrier. Much of the analytic approach is discussed in section 3.2.3. Within the program, selection of the potential is made from the **Potential** item in the main menu. The step potential and square barrier are explicitly listed as choices; the square well is obtained by selecting the square barrier and assigning a negative height in the parameter input screen.

3.2 Double Step

Compare scattering from a step to scattering from a double step of the same total height. Adjust the intermediate step to eliminate reflection. Both potentials are available in the program as the choice **Step** under the **Potential** menu item. In fact, the step potential is treated as a special case of the double step where the height of the intermediate step is set equal to V_∞.

3.3 Smooth Step

Compare scattering from a step to scattering from a ramp potential, and from a smooth step $V(x) = V_0/[1 + \exp(-x/a)]$. Obtain plots of the transmission and reflection probabilities over an energy range from V_0 to at least $2V_0$. Each of these potentials is available in the program as a separate choice under **Potential**.

3.4 Delta Function Barrier

Consider reflection from high, narrow potentials, such as square and Gaussian barriers, for which $g \equiv \int_{-\infty}^{\infty} V(x)dx$ is constant. Compare results with the analytic result for $V(x) = g\delta(x)$, which yields a reflection probability of $R = 1/[1 + 2\hbar^2 E/mg^2]$. Study the wave functions and verify the emergence of a discontinuity in the slope as the delta function shape is approached.

3.5 Dimensional Analysis

Verify that scaled results obtained for a square well depend only on the value of ζ. For a discussion of scaled quantities, see section 3.3.1; ζ is defined there in Eq. 3.41.

3.6 Electron Scattering

A simple model for electron scattering from a negatively ionized gas atom is a square barrier of height 10 eV and thickness 0.18 nm. Compute the transmission probability as a function of energy. If a 20 mA beam of 5 eV electrons strikes the barrier, how much current will get through?

3.7 Ramsauer-Townsend Effect

There exists an attractive potential between an electron and an atom of an inert gas. It can be modeled by a square well of width 0.2 nm. Experiments first conducted by Ramsauer show that electrons near 0.7 eV in energy are almost never reflected. What depth results in nearly zero reflection at this energy? Consider the effect of a slight smoothing of the edges, by use of the Woods-Saxon potential, $V(x) = V_0/[1 + \exp\{(|x| - c)/a\}]$; values of 0.01 nm and 0.1 nm for a and c, respectively, are reasonable choices.

3.8 Square Well Resonances

a. Compare the energies at which the transmission probability for a square well is peaked with those at which bound states are found in an infinite well of the same width.

b. Study the wave functions for these resonances. How many nodes are found inside the well?

c. Compare the behavior of time-dependent wave packets, which can be observed using the program **QMTime**; measure the time spent in the well and compare this with the magnitude of the transmission probability.

d. Measure the spacing between the resonances and graph this versus the dimensionless parameter ζ. (See Eq. 3.41 for the definition of ζ.) Does the spacing increase or decrease?

e. Graph the width of the peaks in the transmission probability versus energy. Are the peaks broader at higher energies?

3.9   Double-Barrier Resonances

Arrange two square barriers such that each individual barrier is narrow but strongly reflective. The choice of **Double Square** under the **Potential** menu item provides a potential that can be adjusted to this form. For a narrow individual barrier to be strongly reflective, it must be tall.

a. Find analytically the energies at which bound states would exist between the barriers if the barriers were completely reflective; this is, of course, the situation of an infinite square well. Use the program to study the transmission and reflection probabilities for energies near these bound state energies, and thereby confirm the existence of resonances, seen as sharp peaks in the transmission probability.

b. Study the wave functions at and away from these resonances. How many nodes are present between the barriers? Does the number of nodes change from one resonance to the next? Compare them with bound state wave functions obtained analytically or computed with **Bound1D**.

c. Compare this stationary-state analysis with the behavior of wave packets computed by **QMTime**, particularly with regard to the time spent by the packet between the barriers.

3.10   Double Delta Function Resonances

Show that transmission through a double delta function barrier $\lambda\delta(x - a) + \lambda\delta(x + a)$ exhibits resonances. See Exercise 3.4 for a brief discussion of how to model a single delta function barrier. This exercise can then be considered a limiting case of Exercise 3.9.

3.11   Alpha Decay

Model alpha decay, in which the alpha particle must tunnel through a Coulomb barrier. Assume a potential of the form

$$V(x) = \begin{cases} -50 \text{ MeV} & , x < 10 \text{ fm} \\ 16 \text{ MeV} & , 10 \text{ fm} < x < 20 \text{ fm} \\ 0 & , x > 20 \text{ fm} \end{cases}$$

After tunneling, the alpha's energy is 5 MeV. Compute the transmission probability.

As required by the program, V must be shifted up to keep $V = 0$ for $x < x_{min}$. Since the program is written for electron scattering, a rescaling of input and output must also be done. A natural energy scale is $V_0 = 50$ MeV and a natural length scale is $L_0 = 10$ fm. A value of ζ can be computed from the definition (Eq. 3.41), with m the mass of the alpha particle. The scaled Schrödinger equation (Eq. 3.40) is then identical to that of any electron problem in which ζ has the same value, and $\overline{V}(\overline{x}) = [V(\overline{x}L_0) + 50 \text{ MeV}]/V_0$ has the same shape. The length scale of the electron problem is fixed at 1 nm in the program; therefore, any distances passed to or from the program should be multiplied or divided, respectively, by 10 fm/ 1 nm = 10^{-5}. The energy scale of the electron problem is given by $V_{oe} = \hbar^2\zeta/(2m_e \times 1 \text{ nm}^2)$, with m_e the electron mass. Any energy passed to or from the program should be multiplied or divided by 50 MeV/$V_{oe} = 1312.3 \ \zeta^{-1}$.

3.12 **Fission Neutrons**
Study the probability of reflection from a nucleus for 5 MeV neutrons. Model the interaction as a step down to a potential energy of −50 MeV. Does rounding of the step reduce or increase the reflection? (A rounded step can be obtained by selecting **Smooth Step** as the choice of potential.) Because the program treats electron scattering, the input and output must be properly scaled. For a discussion of scaling, see Exercise 3.11.

3.13 **Fusion**
Model the fusion of protons to carbon nuclei as a tunneling of the protons through a one-dimensional barrier. The shape of the barrier is like that for alpha decay (see Exercise 3.11.), but the proton approaches from the opposite side. The typical proton energy is determined by the energy distribution associated with a star's temperature, and can be estimated to be no more than a few multiples of kT, where the temperature T is on the order of 10^7 K. Use the program to obtain the transmission probability for energies in this range.

Projects

3.14 **Cold Emission**
If an electric field is applied to the surface of a metal, electrons may be drawn out at room temperature. The model potential is

$$V(x) = \begin{cases} 0 & , x < 0 \\ V_0 - e\mathscr{E}x & , x > 0, \end{cases}$$

where V_0 is the sum of the work function and the Fermi energy of the metal, and $\mathscr{E}$ is the field strength. Add this potential to the program, and find the transmission probability for electrons of various energies below the Fermi level.[10] Compare results with those obtained using the Wentzel-Kramers-Brillouin (WKB) approximation, or with an analytic solution based on Airy functions.[11]

3.15   **Conduction Bands**

The motion of conduction electrons in a metal can be modeled by studying their transmission through a series of potential barriers.[12] The conduction bands of the model will be associated with high transmission probability. Add such a potential to the program; the double square barrier potential already included can serve as an example to follow. Compute the transmission probability for energies up to the tops of the barriers. Compare the ranges of energies at which transmission is likely with the energy bands obtained in a bound state analysis done using the Kronig-Penney program **Latce1D** described in chapter 5.

3.16   **Alpha Decay Through a Coulomb Barrier**

The square barrier model for alpha decay that is suggested in Exercise 3.11 can be improved by use of a Coulomb barrier. This requires addition of a new potential to the program. Also, the input and output are simplified if the program is modified to use the scales and mass appropriate to the problem. Carry out these modifications and repeat the calculations of the earlier exercise. Compare results with those of Exercise 3.11 and with those obtained in a WKB approximation.

3.5   *Details of the Program*

3.5.1   Running the Program

The primary means for control of the program is a set of menu options. These options are as follows:

- **File:** - Get program information; read and write data files; exit the program.

  - **About CUPS:** Show description of software consortium.

  - **About Program:** Show credits and a brief description.

  - **Configuration:** Verify and/or change program configuration.

  - **New:** Set file name to default file name and start new calculation.

  - **Open...:** Open file and read contents.

  - **Save:** Save current state of the program to a file.

  - **Save As...:** Save current state of the program to a file with chosen name and set file name to this choice.

  - **Exit Program**

- **Parameters:** Change numerical parameters and the choice of algorithm.

- **Potential:** Display, choose, and modify the potential energy.

 – **Display & Modify Current Choice:** Plot current choice for potential and display parameter values; allow modification.

 – **Choose & Modify: Square Barrier**

  Double Square

  Step

  Ramp

  Smooth Step

  Gaussian

  Woods-Saxon

  User Defined

 Choose this potential; plot and display parameter values; allow modification.

• **Compute:** Compute and plot results.

 – **Wave Function:** Plot previously calculated wave function, if any; then compute and plot new results.

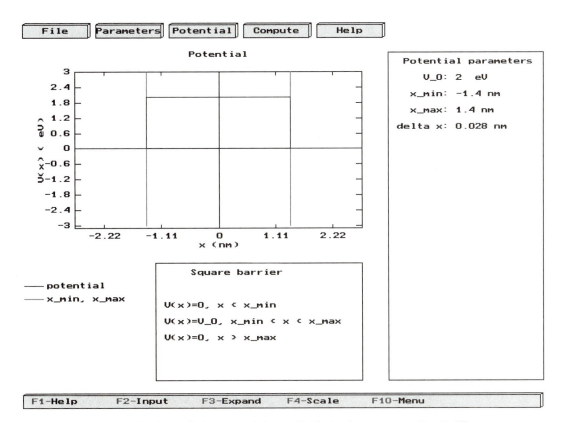

Figure 3.2: One of the potentials available in the program **Scattr1D**.

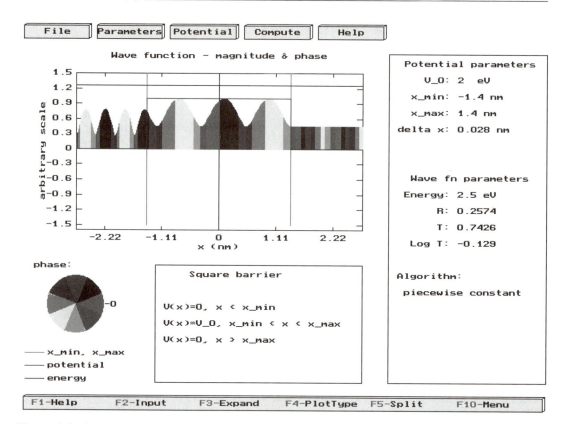

Figure 3.3: A stationary state wave function for the square barrier. This particular kind of plot, which is one of four possibilities, uses color to represent the phase of the wave function.

> – **Trans & Refl Probabilities:** Plot previous results for transmission and reflection probabilities versus energy, if any; then compute and plot new results.

- **Help:** Display help screens.

> – **Summary:** Display summary of menu choices.
>
> – **'File':** Describe entries under **File**.
>
> – **'Parameters':** Describe input of parameters.
>
> – **'Potential':** Describe entries under **Potential**.
>
> – **'Compute':** Describe entries under **Compute**.
>
> – **Algorithms:** Describe key algorithms used in program

Within options under **Potential** and **Compute**, input screens for parameter values are presented. Also within these options, certain keys are activated for control of the display and of calculations. Some keys bring up input screens that request additional information, such as choice of a display type for the wave function.

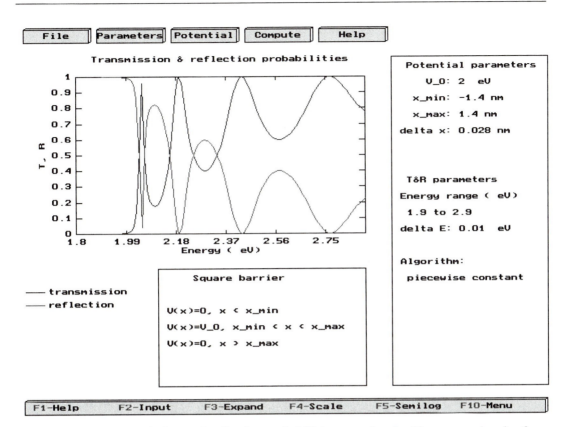

Figure 3.4: The transmission and reflection probabilities associated with a square barrier for a range of beam energies.

3.5.2 Sample Output

Figures 3.2–3.4 illustrate some of the output that can be obtained with the program **Scattr1D**. The appearance of the plots is much enhanced when a color display is available.

References

1. Merzbacher, E. *Quantum Mechanics,* 2nd ed. New York: Wiley, 1970, p. 80.

2. Anderson, E. E. *Modern Physics and Quantum Mechanics.* Philadelphia: Saunders, 1971, p. 158.

3. Gasiorowicz, S. *Quantum Physics.* New York: Wiley, 1974, p. 75.

4. Brandt, S., Dahmen, H. D. *The Picture Book of Quantum Mechanics.* New York: Wiley, 1985, p. 75.

5. Eisberg, R., Resnick, R. *Quantum Physics of Atoms, Molecules, Solids, Nuclei and Particles,* 2nd ed. New York: Wiley, 1985, p. 184.

6. Liboff, R. L. *Introductory Quantum Mechanics,* 2nd ed. San Francisco: Holden-Day, 1992, p. 220.

7. Park, D. A. *Introduction to the Quantum Theory,* 3rd ed. New York: McGraw-Hill, 1992, p. 94.

8. Kalotas, T. M., Lee, A. R. "A New Approach to One-Dimensional Scattering," American Journal of Physics **59**(1):48, 1991.

9. Gordon, R. G. "New Method for Constructing Wave Functions for Bound States and Scattering," Journal of Chemical Physics **51**(1):14, 1969. "Quantum Scattering Using Piecewise Analytic Solutions," Methods in Computational Physics **10**:81, 1971. These contain careful discussion of improvements for accuracy, particularly for a linear approximation to the potential. In addition, see the caveat in Senn, P. "Numerical Solutions of the Schrödinger Equation," American Journal of Physics **60**(9):776, 1992.

10. Greenhow, R. C., Matthew, J. A. D. "Continuum computer solutions of the Schrödinger equation," American Journal of Physics **60**(7):655, 1992.

11. Abramowitz, M., Stegun, I. A. (eds) *Handbook of Mathematical Functions.* New York: Dover, 1964.

12. Walker, J. S., Gathright, J. "A Transfer-Matrix Approach to One-Dimensional Quantum Mechanics Using Mathematica," Computers in Physics **6**(4):393, 1992.

4

Quantum Mechanical Time Development

Daniel F. Styer

4.1   Introduction

Think of a typical introductory physics problem: "A rock is tossed horizontally off of a 100 meter high cliff at speed 3 meters/second. How far from the base of the cliff does it land?" This is a time development problem. "Find the period of a satellite orbiting the earth at an altitude of 500 km." Another time development problem. $\mathbf{F} = m\mathbf{a}$ is a time development equation. Indeed, time development is at the very center of classical mechanics.

It is surprising, then, that in quantum mechanics one rarely solves a time development problem. You are flooded with problems involving energy eigenvalues, reflection coefficients, and uncertainty relations, but you are hardly ever presented with a problem of the familiar form "This is the situation now. What will it be 7 seconds from now?" There are two reasons for this. First, it is difficult to detect quantal time development experimentally. Even a century ago it was easy to measure the energy eigenvalues of an electron in an atom through its emission spectrum, but the corresponding motion of an electron around a nucleus was first detected experimentally by Yeazell and Stroud[1] in 1988. Second, the analytical solution of quantal time-development problems is mathematically intricate. In many simple cases no analytic solution has ever been uncovered. This difficulty is becoming less severe as computers become widespread and numerical solutions are used when analytic solutions cannot be found . . . or even when they can. The aim of program **QMTime** is to harness this trend in order to render problems in quantal time development accessible to physics students on an everyday basis.

4.2   *Analytic Approach to Time Development*

Throughout this chapter we consider the nonrelativistic motion of a single particle in one dimension, subject to a potential energy function $V(x, t)$. (In most cases the potential energy function will be independent of time. One common exception arises when a charged particle is subjected to the oscillating electric field of a light wave.) In *classical mechanics* the time development problem consists of solving the ordinary differential equation

$$m \frac{d^2 x(t)}{dt^2} = - \frac{\partial V(x, t)}{\partial x} \tag{4.1}$$

for $x(t)$ subject to the initial conditions

$$x(0) = x_0, \quad \frac{dx}{dt}(0) = v_0. \tag{4.2}$$

In *quantum mechanics* the time development problem has an entirely different form. It consists of solving the partial differential equation

$$i\hbar \frac{\partial \Psi(x, t)}{\partial t} = - \frac{\hbar^2}{2m} \frac{\partial^2 \Psi(x, t)}{\partial x^2} + V(x, t)\Psi(x, t) \tag{4.3}$$

for the wave function $\Psi(x, t)$ subject to the initial condition

$$\Psi(x, 0) = \Psi_0(x). \tag{4.4}$$

The time development equation (Eq. 4.3) is called the Schrödinger equation or the time-dependent Schrödinger equation. (Schrödinger himself[2] called it the "real" wave equation to distinguish it from the energy eigenequation, or time-independent Schrödinger equation, which is less fundamental but which he discovered first.)

   Sometimes it is possible to solve the time development problem analytically (see Exercises 4.1 and 4.2) but such solutions are restricted to certain special potentials with certain particular initial conditions. There is an important approximate technique, the method of stationary phase,[3-6] which gives useful partial information under some (not all) circumstances, but to treat the problem in full generality one must use numerical techniques.

4.1   Exercise: Free Gaussian Wave Packet

Suppose that the potential energy function vanishes and that the initial wave function is a "Gaussian wave packet"

$$\Psi_0(x) = \frac{1}{\pi^{1/4} \sqrt{\sigma}} \exp\left(- \frac{[x - x_0]^2}{2\sigma^2} \right) e^{ip_0 x/\hbar}. \tag{4.5}$$

Verify that the initial value problem (Eqs. 4.3, 4.4) is solved by

$$\Psi(x, t) = \frac{1}{\pi^{1/4} \sqrt{\sigma[1 + i\frac{\hbar}{m\sigma^2}t]}} \exp\left(- \frac{[x - (x_0 + \frac{p_0}{m}t)]^2}{2\sigma^2[1 + i\frac{\hbar}{m\sigma^2}t]} \right) e^{i(p_0 x - E(p_0)t)/\hbar}, \tag{4.6}$$

where $E(p_0) = p_0^2/2m$. (Warning: This is not easy! It helps to define $\alpha(t) = 1 + i\frac{\hbar}{m\sigma^2}t$.)

4.2 **Exercise: Characteristics of a Free Gaussian Wave Packet**
Deduce the following from expression 4.6:

a. $\langle x(t) \rangle = x_0 + \frac{p_0}{m} t.$

b. $\Delta x(t) = (\sigma/\sqrt{2}) \sqrt{1 + (\frac{\hbar}{m\sigma^2} t)^2}.$

c. $\langle p(t) \rangle = p_0.$

d. $\Delta p(t) = \hbar/(\sqrt{2}\sigma).$

e. $\langle E(t) \rangle = p_0^2/(2m).$

f. The probability density $|\Psi(x, t)|^2$ is symmetric about $\langle x(t) \rangle.$

4.3 *Computational Approach to Time Development*

It is important to realize that any computer program is only a *simulation* of the real world, not a perfect reproduction of it. (Indeed, the same is true of any theoretical model.) In our case, for example, a perfect mimic of time development would have to start by storing the initial wave function $\Psi_0(x)$. Because x is continuous, this would require infinite memory. No computer can (or ever will be able to) perform even this simple preliminary task.

Instead, the value of the wave function is stored only at a finite number of grid points, separated by a distance Δx and running from $x = x_{\min}$ to $x = x_{\max}$. (In program **QMTime**, the values of these parameters are 0.0469 nanometer, -6 nm, and 6 nm $- \Delta x$, respectively. The value of Δx was selected to admit exactly $256 = 2^8$ grid points; an integral power of 2 is needed in order to find the momentum wave function.) The wave function is assumed to vanish for $x \leq x_{\min}$ and for $x \geq x_{\max}$. This is equivalent to placing infinitely high potential energy barriers ("edge walls") at $x_{\min}$ and $x_{\max}$. This "spatial discretization" approximation is a good one whenever the magnitude of the change in the wave function from one grid point to the next is small compared with the magnitude of the wave function in that vicinity (see Exercise 4.33).

In addition to the spatial discretization, there must be a discretization in time as well. We will consider time steps of magnitude Δt which, for accuracy, must be chosen so that the magnitude of $\Psi(x, t + \Delta t) - \Psi(x, t)$ at any grid point is small compared with the magnitude of $\Psi(x, t)$ in that vicinity. The value used by **QMTime** is 0.025 femtosecond (fs).

To propagate the discretized wave function forward by a discrete amount of time, we must use a discretized form of the Schrödinger equation. Several discretization schemes are available: the one used here, called Crank-Nicholson, is very reliable. Section 4.7 discusses this scheme in detail, points out some of its attractive features, and mentions some of the alternatives.

4.4 *Tour of Program* QMTime

This section describes the highlights of **QMTime**; before reading it, you should read appendix B "The Display of Wave Functions." More detail is given in section 4.5, "Running Program **QMTime** (Reference)."

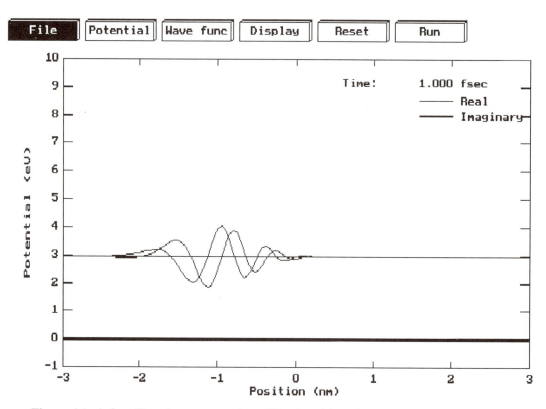

Figure 4.1: A free Gaussian wave packet. ("Real and imaginary parts" display style.)

4.4.1   Free Particle

When the program starts up it displays a welcoming message, then settles down into the status that it will maintain for as long as it is running: namely, displaying the current wave function and potential energy function. The default potential energy function is flat, $V(x) = 0$, and the default initial wave function is a Gaussian wave packet. Choose menu item **Run** to see how this initial wave function changes with time. (See Figure 4.1. Because the default "color for phase" display style is captured poorly by black-and-white illustrations, the figures in this chapter are drawn using the "real and imaginary parts" display style.)

Notice that the screen shows only the middle half of the allowed position values, which range from $x_{min} = -6$ nm to $x_{max} \approx +6$ nm. Thus a wave function can move off of the screen without difficulty. In contrast, the simulation halts if the wave function approaches the edge walls at ± 6 nm, because otherwise the display would show spurious effects due to reflection from the infinitely high edge walls. If you want to see a simulation again, choose **Reset this Run** from the **Reset** menu.

4.3   Exercise: Motion of a Free Gaussian Wave Packet

a. Check to see that the features mentioned in items a, b, e, and f of Exercise 4.2 are satisfied qualitatively.

b. Choose **Things to Notice** from the **File** menu and verify the qualitative features pointed out there.

c. Can you uncover other features of the motion that are clear from the computer screen but that are not readily apparent from the analytic solution (Eq. 4.6)?

One of the advantages of computer solution of quantal time development problems is that, once the program is written, it is easy to apply it to any initial wave function. Choose **Lorentzian** from the **Wave func** menu for a different initial wave packet, then see how it changes with time. In what respects are the motions of Lorentzian and Gaussian wave packets similar, and in what respects are they different?

4.4.2 Free Particle in the Momentum Representation

The position space wave function $\Psi(x, t)$ contains all the information available concerning the state of the system. But $\Psi(x, t)$ is not the only such function. The *momentum space wave function*

$$\tilde{\Psi}(p, t) = \frac{1}{\sqrt{2\pi\hbar}} \int_{-\infty}^{+\infty} \Psi(x, t) e^{-i(p/\hbar)x} \, dx \qquad (4.7)$$

also contains all the information.

Choose **Display | Plot What** and check the boxes to show both position and momentum wave functions. How do you think the momentum wave function will change with time? Choose **Run** to see whether your guess is correct (see Figure 4.2). When you are finished, go back to **Display | Plot What** and remove the momentum wave function display.

4.4.3 Changing the Potential Energy Function

Choose **Potential | Two Square Barriers** to change the potential energy function. Notice that this choice does *not* change the wave function: Whatever wave function was current remains current after the potential energy is changed, even if it is no longer a particularly appropriate wave function. (The mean energy of the wave function will probably change, and if it does the graph of the wave function will shift up or down.) This feature enables **QMTime** to explore the fate of a wave function as the potential changes with time, either slowly or rapidly (the so called "adiabatic or instantaneous change" problem), and it is also needed for simple consistency: A choice from the **Wave func** menu does not change the potential, and a choice from the **Potential** menu does not change the wave function. The only exception to this rule comes if the new potential is at some places infinite, as in the infinite square well. In this situation non-zero wave function outside of the well is impossible, not just inappropriate, so the existing wave function is clipped to fit within the well and renormalized.

Choose **Wave func | Gaussian** and accept the defaults for a wave function more appropriate to the problem of two square barriers, then choose **Run** to try it out. (See Figure 4.3.) How does this time evolution differ from that in an infinite square well of the same width?

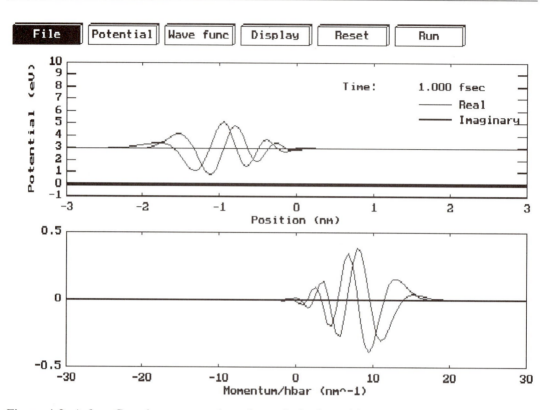

Figure 4.2: A free Gaussian wave packet, shown in both position space and momentum space representations. ("Real and imaginary parts" display style.)

4.5   Running Program QMTime (Reference)

4.5.1   Menu Items

- **File**: Documentation, control, and bookkeeping items.

  - **About CUPS**: Information about the CUPS project.

  - **About Program**: Information about program **QMTime** and its display.

  - **Things to Notice**: Points out features of free particle motion.

  - **Configuration**: Modify colors, temporary directories, and other program characteristics.

  - **Open...**: Read a profile file and set the current profile accordingly. See the description of **Reset | Reset this Run** in this section for more information about profiles.

  - **Save as...**: Save the current profile information in a file.

  - **Exit Program**: Leave the program.

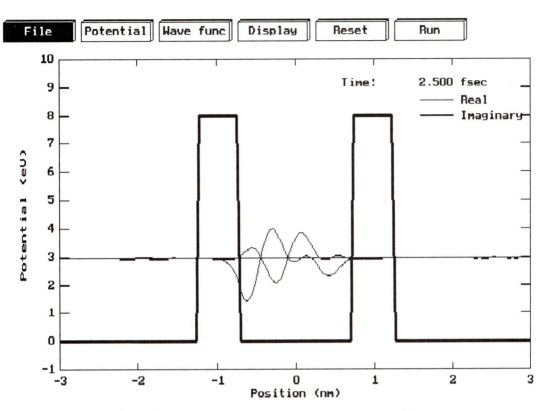

Figure 4.3: A wave packet between two square barriers.

- **Potential**: Set the potential energy function.

  – *Various named potentials*: Set the potential energy function to the potential named. Most of these entries, once chosen, will pop up an input screen that asks for parameters concerning the potential energy function.

  – **User Defined**: If you want to study a time-independent potential that is not on the list, you may code it into the **Pascal** function **UserDefinedPotential**. In the original program, this produces a simple V-shaped potential, but you may modify the procedure and recompile the program to get whatever you want.

  – *Note*: For the most part, any of the above choices from menu **Potential** will change the potential only and not the wave function (see section 4.4.3). There is an exception if the new potential has infinite values where the wave function is non-zero. (This can occur when changing to the infinite square well, to the double square well, or possibly to a user defined potential.) In this case the wave function is set to zero wherever the new potential is infinite, and the surviving wave function is multiplied by a real constant to preserve normalization.

  – **Driving**: Introduce a driving force of the form $F(t) = F \sin(\omega t)$. This is the only way in which **QMTime** can make the potential energy function depend on time.

- **Wave func**: Set the initial wave function (upper entries) or alter the current wave function (lower entries).

  - **Gaussian**: Set the initial wave function to a Gaussian wave packet

    $$\Psi_0(x) = \frac{1}{\pi^{1/4}\sqrt{\sigma}}e^{-(x-x_0)^2/2\sigma^2}e^{ip_0x/\hbar}. \qquad (4.8)$$

    An input screen pops up asking you to enter the mean values of the position and energy, and the uncertainty in position. From this information the program calculates the parameters x_0, σ, and p_0 (see Exercise 4.2).

  - **Lorentzian**: Set the initial wave function to a Lorentzian wave packet

    $$\Psi_0(x) = \frac{A^2}{(x-x_0)^2 + \gamma^2}e^{ip_0x/\hbar}, \qquad (4.9)$$

    where normalization fixes $A^4 = 2\gamma^3/\pi$. An input screen pops up asking you to enter the mean values of the position and energy, and the uncertainty in position. From this information the program calculates the parameters x_0, γ, and p_0 (see Exercise 4.35).

  - **Tent**: Set the initial wave function to a symmetric triangular wave packet of height H

    $$\Psi_0(x) = \begin{cases} 0 & \text{for} & (x-x_0) & \leq -L \\ \frac{H}{L}(L+(x-x_0))e^{ip_0x/\hbar} & \text{for} & -L \leq (x-x_0) & \leq 0 \\ \frac{H}{L}(L-(x-x_0))e^{ip_0x/\hbar} & \text{for} & 0 \leq (x-x_0) & \leq L \\ 0 & \text{for} & L \leq (x-x_0), \end{cases} \qquad (4.10)$$

    where normalization fixes $H^2 = 3/(2L)$. An input screen pops up asking you to enter the mean values of the position and energy, and the uncertainty in position. From this information the program calculates the parameters x_0, L, and p_0 (see Exercise 4.36).

  - **Combination of Energy States**: This entry, which is active only for the potentials infinite square well or harmonic oscillator, allows you to choose as initial wave function any linear combination of the three lowest-lying energy eigenfunctions.

  - **User Defined**: If you want to study an initial wave function that is not built in, you may code it yourself into **QMTime**'s procedure **UserDefinedWF**. In the original program, this produces a simple, pure real Gaussian wave packet, but you may modify the procedure and recompile the program to get whatever you want.

  - **Shift**: Alter the current wave function by translating it a given distance (which is read from the screen).

  - **Stretch**: Stretch the current wave function about its mean position. An input screen pops up asking you to enter the stretch factor. For a compression, use a stretch factor less than one.

– **Reflect**: Reflect the current wave function about its mean position.

– **Conjugate**: Take the complex conjugate of the current wave function.

● **Display**: Set what the program displays and how it displays it.

– **Plot How**: Set the display style: color for phase,[7] real and imaginary parts, or probability density. (See appendix B for a description of these display styles.)

– **Plot What**: The display can include any combination of the position wave function, the momentum wave function, and—for the infinite square well and harmonic oscillator potentials—the energy spectrum. It can even display nothing at all, which is useful if you want to do a long calculation quickly, seeing only the final and not the intermediate wave functions. (**QMTime** spends only about 3% of its time actually calculating wave functions...the rest is spent updating the display.)

– **Show Mean Values**: Show vertical lines marking the expectation value of position, $\langle x(t) \rangle$, and of momentum, $\langle p(t) \rangle$. If the lines are present, they can be turned off by choosing **Show Mean Values** a second time.

– **Show Uncertainties**: Show horizontal lines representing the position and momentum uncertainties. Each line is centered on the mean value and is twice as long as the uncertainty, so that the position line, for example, stretches from $\langle x(t) \rangle - \Delta x$ to $\langle x(t) \rangle + \Delta x$. The numerical values are also shown. These displays can be turned off by choosing **Show Uncertainties** a second time.

● **Reset**.

– **Reset this Run**: Go back to the current initial wave function with the current potential energy and the current display and animation speed settings. This will *not* reproduce the most recent run if the potential energy was changed after the initial wave function was set.

– **Reset to Defaults**: Go back to the situation in which the program starts up.

● **Run**: Start the time evolution.

4.5.2 Hot Keys

While a simulation is evolving in time, six hot keys become active. When the simulation is stopped (either through a collision with the edge walls, or by clicking a menu item, or by pressing the hot key **F10-Menu**) the hot keys go away.

● **F2-Stop** or **F2-Run**: When a simulation is started (by choosing **Run** from the menu) the animation runs continuously. You may stop it (freeze the frame) by pressing key **F2-Stop**. Once the animation is frozen the label for key **F2** becomes **F2-Run**. You may start the animation running continuously again either by pressing key **F2-Run** or by choosing any menu item.

- **F3-Step**: While a simulation is stopped, it can be advanced one frame at a time by pressing key **F3-Step** or by clicking the mouse.

- **F4-Backward** or **F4-Forward**: The simulation may run either forward or backward in time. The backward possibility is particularly useful if some interesting feature slipped by while you weren't watching carefully. If the simulation is running forward in time, press key **F4-Backward** to reverse the time flow, and if it is running backward, press key **F4-Forward** to start time flowing forward again. The only way to change the direction of time flow is by pressing key **F4**. If time is flowing backward, and you stop the run and use the menu to pick out a completely different simulation, then when you choose **Run** again the simulation will still be going backward in time. (If you don't like this feature, then look in procedure **ProcessMenu** for instructions on changing it.)

- **F5-Slower** and **F6-Faster**: Adjust the animation speed. As with key **F4**, changes made here are not altered through choices from the menu. These keys affect only the presentation made by the program and not the underlying computation. (For example, pressing **F6-Faster** once does not increase the time step for the computation, which would decrease its accuracy. Instead it causes the program to display the evolving wave function only after every other step. Because the program spends only about 3% of its time calculating the wave function, and the rest updating its display, this results in a considerable speed up.)

- **F10-Menu**: Stop the simulation and go to the menu. Useful on mouseless computers, but users with mice will usually just click on the desired menu item for the same effect.

4.5.3   Tips

- The particle in program **QMTime** has the mass of an electron. (It would, however, be slightly incorrect to say that the particle *is* an electron, because this simulation ignores spin.)

- When **QMTime** starts up, it looks in the current working directory for a profile file named **QMTInit.pfl**. If it exists, the program reads that file and sets the profile accordingly. This is a good way to set up the program to suit your individual tastes. For more information on profiles and profile files, see the description of menu items **File | Save as . . .** and **Reset | Reset this Run**.

- Because of limitations in the graphics utilities used by **QMTime**, potential energy functions that actually have vertical lines, e.g., the square wells, show up instead with lines that slope steeply over an interval of width Δx. In such cases the actual vertical lines should be located at the point of *higher* potential energy.

- In the color for phase display style, the program finds the color boundary by interpolating between grid points. Usually this gives the correct result, but look closely at, for example, the second excited state of the infinite square well.

- Most of the time the zero axis for the wave function in position space is a thin blue line positioned at the mean energy. If the mean energy is too high or too low to fit on the graph, then the axis turns red and is positioned at zero energy.

- The presence of a working mouse can slow **QMTime** by 50%. You may wish to turn your mouse off: simply refrain from loading the mouse driver.

4.6   Exercises

4.6.1   Running QMTime

QMTime opens innumerable paths for exploration. Here are some of them.

4.4   **Stability of Time Development Algorithm**
When any time development simulation (classical or quantal) is run forward for a given amount of time, and then backward for the same amount of time, it should recover the initial state. (This is often the first test one makes in debugging a time development simulation. Be aware, however, that it is possible for the simulation to pass this test yet still be faulty!) Set the potential to infinite square well and the initial wave function to the default Gaussian wave packet. Notice precisely what the initial state looks like. Run the simulation forward for a good long time, say 20 fs, then reverse it using key **F4-Backward**. As far as your eye can determine, is the initial wave packet recovered?

4.5   **Time Development of Energy Eigenstates**
An energy eigenfunction $u_n(x)$ with energy E_n evolves in time as $u_n(x)e^{-iE_nt/\hbar}$. Test this analytic result for the ground state and first two excited states of the harmonic oscillator, by changing the potential to harmonic oscillator and then using **Wave func I Combination of Energy States** to set the system into a pure energy eigenstate.

a. In what sequence do the colors change as time evolves? Why?

b. Show analytically that the wave function changes in time periodically with period h/E_n.

c. Use the program to measure how long it takes to make one entire traversal of the color wheel and compare with the analytic value. (Read the value of E_n directly from the screen by noting the height of the blue $\Psi(x, t) = 0$ axis.)

The Free Particle

4.6   **Group Velocity and Phase Velocity**
The "group velocity" of a wave packet is the speed at which the hump moves. The "phase velocity" is the speed at which the boundaries between the colors move. According to the method of stationary phase, the group velocity of a free quantal wave function is exactly twice its phase velocity.

a. Test this prediction for free Gaussian wave packets with position uncertainties of 0.2, 0.4, 0.6, and 0.8 nm. Is the prediction more accurately obeyed for narrow or wide wave packets?

b. Watch the motion carefully and notice that not all color boundaries move at the same speed. Is this dispersion of phase velocities greater for narrow or wide wave packets?

c. Use the result of part b to explain the result of part a.

d. How is part b related to the Heisenberg uncertainty principle?

4.7   Dispersion

Notice that as a free wave packet moves, the colors "bunch up" at the front of the wave packet.

a. Show analytically that this behavior is expected for Gaussian wave packets. (Hint: Apply Eq. 4.22 to Eq. 4.6.)

b. If the phase velocity of a wave decreases with wave number k, it is said to exhibit "ordinary" dispersion. If the phase velocity increases with wave number, it exhibits "anomalous" dispersion (see, for example, Elmore and Heald[8]). Which sort of dispersion is displayed by the Schrödinger equation? How does this explain the bunching of colors for all wave packets, not just Gaussians?

4.8   Dispersion in Various Display Styles

In the "color for phase" display style, the effects of dispersion are seen through the bunching of colors at the front of the wave packet. How would this effect appear in the "real and imaginary parts" display style?

4.9   Lorentzian Wave Packets

Some of the features exhibited by the time development of a free Gaussian wave packet are:

i. Uniform motion of $\langle x(t) \rangle$.

ii. Increase of $\Delta x(t)$ with time.

iii. Probability density is always symmetric about $\langle x(t) \rangle$.

iv. As time evolves the momentum space wave function $\tilde{\Psi}(p, t)$ changes in phase but not in magnitude.

v. Phase velocity is less than group velocity (see Exercise 4.6).

vi. Local wavelength decreases toward the front of the packet (dispersion: see Exercise 4.7).

Experiment to determine which of these features are shared by the time development of free Lorentzian wave packets.

4.10   Tent Wave Packets

Perform Exercise 4.9 for tent rather than Lorentzian wave packets.

Static Wave Functions

The main purpose of program **QMTime** is to demonstrate time development. Nevertheless, the program does reveal some interesting facts concerning static wave functions.

4.11 Mean Energy for Energy Eigenstates

a. Set the potential to harmonic oscillator, and use **Wave func | Combination of Energy States** to set the particle into a pure energy eigenstate. Measure the mean energy from the height of the blue axis on the screen and compare it to the theoretical result

$$E_n = (n + \tfrac{1}{2})\hbar\omega_0, \quad \text{where} \quad \omega_0 = \sqrt{K/m}. \tag{4.11}$$

b. Perform this comparison for the infinite square well also.

4.12 Momentum Space Wave Functions
Use **Display | Plot What** to show both the position space and momentum space wave functions. Then change the potential energy function from **Flat** to **Gaussian Barrier**. How does the momentum space wave function change? Why?

4.13 Translated Wave Functions
Use **Display | Plot What** to show both the position space and momentum space wave functions. Set the wave function to a Gaussian wave packet with position uncertainty of 0.1 nm.

a. Use **Wave func | Shift** to move the wave packet's center from -2.0 nm to $+2.0$ nm in steps of 1.0 nm. Describe the corresponding *momentum space* wave functions in terms of the following:

   i. Magnitude.

   ii. Phase at $p = 0$.

   iii. Rate of change of phase with momentum (including sign).

b. Use **Wave func | Gaussian** to produce wave packets centered on -2.0 nm, -1.0 nm, 0.0 nm, $+1.0$ nm, and $+2.0$ nm. Describe the momentum space wave function in each case.

c. Why are the results of these two parts different?

4.14 Minimum Energy Wave Packets
Use **Display | Plot What** to show both the position space and momentum space wave functions. Using **Wave func | Gaussian**, attempt to produce a free wave packet with negative energy. There is of course no such thing, and the program will instead produce a minimum energy wave packet. Try this with several different potentials and wave packet types.

    a. What are the phase characteristics (in position space) of such minimum energy wave packets?

    b. What is the mean momentum in each case?

4.15   Phase and Vanishing Momentum

    a. The previous exercise might lead you to conjecture that "if $\psi(x)$ has constant phase, then $\langle p \rangle = 0$." Prove this conjecture correct analytically.

    b. It might also lead you to conjecture that "if $\langle p \rangle = 0$, then $\psi(x)$ has constant phase." Prove this conjecture incorrect by demonstrating a counterexample. (Hint: Try a tent wave packet in a harmonic oscillator.)

    c. Produce and prove a conjecture similar to the one in part a that links the phase of $\tilde{\psi}(p)$ with $\langle x \rangle$. And, paralleling what you did in part b, dispose of that conjecture's converse.

4.16   Phase and Momentum
Use **Display | Plot What** in order to show both the position space and momentum space wave functions. Using **Wave func | Gaussian**, produce wave packets with mean energies from 1 eV to 13 eV in steps of 2 eV.

    a. Measure the rate of change of phase with position for each wave packet. Specifically, how far down the axis do you need to move in order to get one complete phase rotation, say from red to red again? Convert to a phase rotation rate.

    b. Relate the phase rotation rate to the mean momentum in each case.

    c. All of these wave packets have *positive* mean momentum. How would the phase of the position space wave function vary in a wave packet with negative mean momentum?

Scattering

In addition to these exercises, you should look at Exercises 3.8 and 3.9, which investigate time behavior at scattering resonances.

4.17   Reflection From an Infinite Step
Use **File | Open . . .** to read in the profile file **BOUNCE.PFL** and then run the simulation to see a well-defined wave packet reflecting from an infinitely high potential barrier. During the collision, the wave packet fragments into numerous jagged mounds, each of nearly constant phase, with the phase difference from one mound to the next being almost exactly π radians. Explain this behavior through a model involving (initially) a rightward-moving wave packet located to the left of the step plus a leftward moving "virtual" wave packet located to the right of the step. If the initial wave packet had a mean momentum of p_0, what is the width of each mound?

4.18 Reflection From a Finite Step

When a wave packet with mean energy E collides with a step of height $V > E$, the particle spends some time with the mean position $\langle x(t) \rangle$ actually underneath the step, in the classically forbidden region. According to the method of stationary phase,[6] that time is $\hbar / \sqrt{E(V - E)}$. Test this prediction. In particular, what happens as $V \to E$?

4.19 Reflection From a Step: Phase Relations

Set the potential to **Step**, then choose **Wave func | Gaussian** and accept the default wave packet. Allow the simulation to proceed until the wave packet breaks up into two rather well defined wave packets, one to the right of and one to the left of the step.

a. For each of the two resultant packets, in what sequence does the phase change with position? (In other words, at a fixed time does the phase increase or decrease as position increases?) Explain the difference in light of Exercise 4.16, part c.

b. View the scattering event again in momentum space. After the collision, which packet in momentum space corresponds to which packet in position space?

c. Estimate the mean energies of both the left-hand and the right-hand wave packets. Relate this to the mean overall energy.

d. Estimate the (group) velocities of both the left-hand and the right-hand wave packets.[9]

4.20 Scaling Through Dimensional Analysis

Consider a potential energy function with a characteristic length L_0 (such as a square barrier of width L_0).

a. Verify that the following quantities are dimensionless:

$$\bar{x} = \frac{x}{L_0}, \quad \bar{t} = \frac{t}{mL_0^2/\hbar}, \quad \text{and} \quad \overline{V}(\bar{x}) = \frac{V(\bar{x}L_0)}{\hbar^2/mL_0^2}. \tag{4.12}$$

b. Show that the Schrödinger equation (Eq. 4.3), when written in terms of these dimensionless quantities, becomes

$$i\frac{\partial \Psi(\bar{x}, \bar{t})}{\partial \bar{t}} = -\frac{1}{2}\frac{\partial^2 \Psi(\bar{x}, \bar{t})}{\partial \bar{x}^2} + \overline{V}(\bar{x})\Psi(\bar{x}, \bar{t}). \tag{4.13}$$

This result implies that if all the lengths in a problem (e.g., barrier widths, position uncertainties) were doubled, and all the energies (e.g., barrier heights, particle expectation energies) were quartered, then the exact same process would take place, but it would take four times as long.

c. Set up the default square barrier with its default Gaussian wave packet and run the simulation for 1.7 fs. Sketch the wave function. Now set up a simulation with all lengths doubled and all energies quartered. Run it for an appropriate time and verify that the wave function is (nearly) reproduced.

(This scaling explains why program **QMTime** does not allow several of the potentials to be varied in width . . . because appropriate changes in times and energies have the same effect.)

4.21   Scattering From a Square Well

Set the potential energy function to a square well of depth 4 eV and width 0.5 nm; set the initial wave function to a Gaussian with $\langle x \rangle = -2$ nm, $\Delta x = 0.2$ nm, and $\langle E \rangle = 6$ eV; and adjust the display to show the probability density in both position and momentum space.

a. Describe the time evolution in momentum space.

b. Stop the evolution at time 3.0 fs, and interpret each of the five peaks in the momentum space probability density. Compare your interpretation to that of Good.[10]

c. What happens if the initial wave function is the corresponding Lorentzian?

4.22   Scattering From Notched Barriers

A notch barrier is the same as a square barrier except that the walls have finite rather than infinite slope.[11]

a. Experiment with various notched barriers, always using the default wave function, barrier height, and barrier width, but varying the slope of the wall. Convince yourself that the amount of fine-scale jagged structure in the probability density during scattering decreases as the width of the sloping wall increases.

b. It is clear physically that such structure will cross over from jagged to smooth as the width of the sloping wall exceeds some "characteristic length" of the wave packet. There are three such characteristic lengths: the position uncertainty Δx, the deBroglie wavelength corresponding to the mean momentum $\hbar/\langle p \rangle$, and the deBroglie wavelength corresponding to the mean energy $\hbar/\sqrt{2m\langle E \rangle}$. Which one is the relevant length for this crossover?

Oscillators

4.23   Harmonic Oscillator Periodicity

Regardless of its energy or initial condition, a classical harmonic oscillator exactly repeats its motion with angular frequency $\omega_0 = \sqrt{K/m}$. Test the mean position of the quantal harmonic oscillator for this behavior. Is the period of $\langle x(t) \rangle$ independent of the energy and wave packet type?

4.24   Infinite Square Well Periodicity

a. Find analytically the period as a function of energy for a classical particle in an infinite square well of width L.

b. Show analytically that any wave function, regardless of energy, in the same infinite square well is periodic in time with period

$$\frac{4mL^2}{\hbar^2\pi}.$$ (4.14)

c. Set the potential to **Infinite Square Well**, and turn on **Display | Show Mean Values**. Measure the half-periods (i.e., $\langle x(t)\rangle$ going from zero to zero once, not twice) of initially Gaussian wave packets with the default mean and uncertainty in position, but with mean energies ranging from 1 eV to 13 eV in steps of 2 eV. Compare your results to the classical and quantal formulas.

d. Measure the half-periods of the three different types of initial wave packets (Gaussian, Lorentzian, and tent), all with default parameters.

e. Qualitatively observe the long-time behavior of initially Gaussian wave packets with mean energies of 1, 7, and 13 eV. (Consider about 20 fs to be a long time. Unless you are exceptionally patient, you will want to press **F6-Faster** about five times.) Comment on this behavior in light of the correspondence principle.

f. Comment on the veracity of the numerical time-development algorithm in this situation.

4.25 **Harmonic Oscillator With Zero Momentum Initial States**
(You may want to look over Exercise 4.14 before doing this one.) One way to start a classical harmonic oscillator is to release the particle from rest but away from its equilibrium position. (This is the classic way to start a child's swing, for example.) The analogous action can be performed in program **QMTime** by setting the potential to **Harmonic Oscillator**, then using **Wave func | Gaussian** to request a wave packet with zero energy. There is no such wave packet, so the program responds by giving you the wave packet of lowest energy consistent with the position and uncertainty that you requested.

a. What are the phase characteristics of this wave packet?

b. What does this imply for the mean momentum of the wave packet? (Hint: Use **Display | Plot What** to show both position and momentum wave functions.)

c. Set this wave packet into motion. What are its phase characteristics at the turning points? When does the phase vary most rapidly with position?

d. Observe the wave function in both position and momentum space. When does the phase of the momentum space wave function vary most rapidly and when is it constant?

e. Verify your observations in parts c and d with an analytic calculation. In particular, show that in a harmonic oscillator of angular frequency $\omega_0 = \sqrt{K/m}$,

$$\Psi\left(x, t + \frac{2\pi}{\omega_0}\right) = -\Psi(x, t).$$ (4.15)

4.26   Harmonic Oscillator Coherent States

The *coherent states* of a harmonic oscillator are important in the quantum theory of radiation.[12] By definition, coherent states are the result of displacing the harmonic oscillator ground state. In **QMTime**, you can produce coherent states by beginning with the ground state (choose **Wave func | Combination of Energy States** and pick a combination with no contribution from the excited states) and then displacing it with **Wave func | Shift**.

a. Produce a coherent state and set it into motion. What remarkable property does its probability density show as time evolves?

b. Do the momentum space wave functions show similar inflexibility?

4.27   Rigidly Sliding Harmonic Oscillator States

Demonstrate that the rigidly sliding behavior of the coherent states comes about when *any* energy eigenfunction, not just the ground state, is first translated and then allowed to evolve in time.[13]

4.28   Variation of Wave Packet Uncertainties in the Harmonic Oscillator

Set the potential to harmonic oscillator and the display to probability density. Show position and momentum uncertainties by choosing **Display | Show Uncertainties**. Use the technique described in Exercise 4.25 to produce a minimum energy Gaussian initial wave packet, and evolve this wave packet forward in time for at least one-half period. When is the product $\Delta x \Delta p$ a minimum?

4.29   Variation of Wave Packet Width in the Harmonic Oscillator

Set the potential to harmonic oscillator and display probability density only. Use the technique described in Exercise 4.25 to produce five minimum energy Gaussian initial wave packets, all with mean position -2.0 nm but with position uncertainties ranging from 0.1 nm to 0.5 nm in steps of 0.1 nm. Evolve each wave packet forward in time for at least one period.

a. For each wave packet, when is its width a maximum? A minimum?

b. Use the method of means[14] to prove that the observed width behavior will result from any pure real initial wave function.

c. Find analytically the position uncertainty of the coherent states described in Exercise 4.26. Compare to the wave packet above with least variation in width.

4.30   Wave Packets in a Quartic Oscillator

The above exercises have demonstrated several features of time development in a harmonic oscillator. For example:

i. An initially Gaussian wave packet always has a probability density symmetric about $\langle x(t) \rangle$.

ii. If the initial wave packet is pure real, a maximum (or minimum) in $\Delta x(t)$ comes when the packet crosses the center point, i.e., when $\langle x(t) \rangle = 0$.

iii. Phase varies with position most rapidly when the wave packet crosses the center point, and least rapidly when it is at a turning point.

Which of these features are shared by time development in a quartic oscillator?

4.31 Sloshing Mode in the Double Square Well
Set the potential to **Double Square Well**, select a minimum energy Gaussian wave packet with the default mean and uncertainty of position, and choose **Display | Show Mean Values** to turn on the display of $\langle x(t) \rangle$. Then choose **Run**.

a. Compare the changes in phase and magnitude to those observed in Exercise 4.5, and argue that this state is "nearly" an energy eigenstate.

It is not, however, exactly an energy eigenstate, and if you wait long enough you will find that probability amplitude seeps out of the left-hand well into the right-hand well. "Long enough" in this case means several hundred femtoseconds, so you will want to press **F6-Faster** about 20 times in order to observe the seepage.

b. What is the phase relation between the part of the wave function in the left-hand well and the part in the right-hand well?

c. How long does it take until $\langle x(t) \rangle = 0$?

(Low energy states in the double square well are well modeled as a two-state system. See, for example, Merzbacher,[4] sections 5.5–5.6, and Feynman et al.,[15] chapters 8 and 9. Other approaches to the double square well are presented in Exercise 2.23 and in reference 16.)

Driving Forces

4.32 Testing Time-Dependent Perturbation Theory
Suppose that a particle starting in the ground state of a harmonic oscillator (of natural frequency ω_0) is exposed to an external driving force $F \sin(\omega t)$. It is natural to ask about the probability that, after being exposed for a time T, it makes a transition to the first excited state. According to time-dependent perturbation theory, that probability is

$$\frac{F^2}{2m\hbar\omega_0} \left(\frac{\sin[(\omega - \omega_0)T/2]}{\omega - \omega_0} \right)^2. \tag{4.16}$$

Use menu options **Potential | Driving** and **Display | Plot What** to test this prediction. Use the default harmonic potential, an exposure time of $T = 1.0$ fs, and driving frequencies ω of 0.2, 0.4, 0.5, 0.6, and 0.66 fs^{-1}.

4.6.2   Dissecting QMTime

Producing a program like **QMTime** involves a number of interesting calculations (both analytical and numerical), and also oceans of mundane "user interface" code. These exercises are designed to demonstrate and explain the interesting sections of the code without forcing you to wade through the oceans. In solving these exercises you might well want to look at the **Pascal** code for **QMTime**, but you will not need to run or to modify the program. There are no exercises here concerning the central time-stepping algorithm because there are many such exercises in section 4.7.

4.33   **When is Spatial Discretization Legitimate?**
Consider a wave packet of the form $A(x)e^{ikx}$ where $A(x)$ varies slowly on the scale of Δx. Show that discretization is legitimate whenever $|k| \ll \pi/\Delta x$. (Hint: What is the smallest wavelength that can be represented on the grid?)

4.34   **Initial Wave Packets**
All of the initial wave packets are of the form

$$A(x)e^{ikx}, \qquad\qquad (4.17)$$

where $A(x)$ is real and independent of the parameter k. (It does, of course, depend on other parameters such as x_0 and σ or γ.)

a. Show that the mean momentum of such a wave packet is $\langle p \rangle = \hbar k$, independent of $A(x)$, while the uncertainty in momentum is

$$(\Delta p)^2 = \hbar^2 \int_{-\infty}^{+\infty} \left[\frac{dA(x)}{dx} \right]^2 dx, \qquad (4.18)$$

independent of k.

b. Show that the mean energy depends upon k through

$$\langle E(k) \rangle = \langle E(k = 0) \rangle + \frac{\hbar^2 k^2}{2m} \qquad (4.19)$$

for any static potential energy function. (This fact is used in procedure **MakeWFPacket**.)

4.35   **Lorentzian Initial Wave Packets**
For the Lorentzian initial wave packets of Eq. 4.9, show that:

a. Normalization requires $A^4 = 2\gamma^3/\pi$.

b. $\langle x \rangle = x_0$.

c. $\Delta x = \gamma$.

d. $\Delta x \Delta p = \hbar/\sqrt{2}$.

4.36   **Tent Initial Wave Packets**
Produce results analogous to those of Exercise 4.35 for the tent wave packet of Eq. 4.10.

4.37 Probability Current Density

Appendix B points out that a wave function can be considered either as real and imaginary parts or as a magnitude and a phase. A third profitable way to view a wave function is as a probability density $|\Psi(x, t)|^2$ and a probability current density

$$j(x, t) = \frac{i\hbar}{2m}\left(\frac{\partial \Psi^*}{\partial x}\Psi - \frac{\partial \Psi}{\partial x}\Psi^*\right). \tag{4.20}$$

(See, for example, Eq. 3.13, or Liboff,[3] pp. 206–208.)

a. In terms of the real functions $A(x)$ (magnitude) and $\phi(x)$ (phase), defined through

$$\Psi(x) = A(x)e^{i\phi(x)}, \tag{4.21}$$

show that

$$j(x) = \frac{\hbar}{m}A^2(x)\frac{d\phi}{dx}. \tag{4.22}$$

b. Demonstrate how, given $|\Psi(x)|^2$ and $j(x)$, you could find $\Psi(x)$ up to an arbitrary overall phase factor.

c. Show that

$$\frac{d\phi}{dx} = \text{Im}\left\{\frac{1}{\Psi}\frac{d\Psi}{dx}\right\}. \tag{4.23}$$

Why is this quantity sometimes called the "local wave number"?

4.38 Other Display Styles

(You should read the previous exercise before working this one.) Evaluate these ideas for displaying complex-valued functions. Will they be appropriate for the display of momentum wave functions or the energy spectrum?

a. Plot $A(x)$ and $\phi(x)$ separately, allowing $\phi(x)$ to vary from 0 to 2π.

b. Plot $A(x)$ and $\phi(x)$ separately, allowing $\phi(x)$ to vary from -10 to $+10$.

c. Plot $A(x)$ and $d\phi(x)/dx$ separately.

d. Plot probability density and probability current density.

e. Invent your own display style (please inform the author!).

4.39 Harmonic Oscillator Energy Eigenfunctions

At two points in the program (procedures **MakeWFCombo** and **FindEnergySpectrum**) we need to know the energy eigenfunctions for the infinite square well and the harmonic oscillator. Those for the infinite square well are easy to compute, because they are simple sines and the sine is a built-in **Pascal** function. Evaluating the harmonic oscillator energy eigenfunctions is more challenging.

a. (Analytic part.) If the harmonic oscillator scaled distance is

$$\xi = \sqrt{\frac{m\omega_0}{\hbar}}x \qquad (4.24)$$

and the eigenfunctions are denoted by

$$u_n(\xi) \quad \text{for } n = 0, 1, 2, 3, \ldots, \qquad (4.25)$$

show that

$$u_n(\xi) = \xi\sqrt{\frac{2}{n}}u_{n-1}(\xi) - \sqrt{\frac{n-1}{n}}u_{n-2}(\xi) \quad \text{for } n = 1, 2, 3, \ldots. \qquad (4.26)$$

Pay special attention to $n = 1$. This recurrence relation is usually used in conjunction with the starting value

$$u_0(\xi) = \frac{1}{\pi^{1/4}}e^{-\xi^2/2}. \qquad (4.27)$$

b. (Numerical part.) The results of part a can be used to evaluate harmonic oscillator energy eigenfunctions only if the recurrence relation (Eq. 4.25) is *stable*. Look up the discussion of stability in Press et al.,[17] section 5.4, and test this recurrence for stability.

4.6.3   Modifying QMTime

Program **QMTime** was intended to be easy to modify. The purpose and action of each subroutine is spelled out by comments at its head. (In particular, these comments state whether and how the subroutine alters global variables.) And at the beginning of the program is a "table of contents" that lists and categorizes all of the subroutines. The following exercises encourage you to explore and modify the code.

4.40   **Input Energies**
If you ask **QMTime** for a wave packet with mean energy 3 eV, you get just about what you ask for. But if you ask for a wave packet with mean energy 8 eV, the resulting wave packet is somewhat deficient in energy. What goes wrong? (Hint: Try increasing the constant **NPoints**.)

4.41   **Scattering From Square Barriers and Square Wells**
The very first animations of time development in quantum mechanics were generated by Goldberg, Schey, and Schwartz[21] in 1967. Their classic paper includes six figures, extracted from movies that they had produced, showing Gaussian wave packets being scattered by square barriers and square wells. Program **QMTime** is distributed with six profile files, **GSS1.PFL** through **GSS6.PFL**, which, when read into the program through **File | Open...**, will reproduce these six historic figures. However, some of the features of the originals are nearly invisible in the reproductions. Make the following changes in **QMTime**, recompile the program, and then watch the reproduced animations in magnified detail.

    i. Change the constant **NPoints** to 400.

    ii. Change the variable **scale** in procedure **PlotWFandPotl** to 6.0.

    iii. Remove the call to procedure **HaltOnEdgeWallCollision**.

4.42 Time-Independent Potentials

Use the "user defined potential" feature to investigate these potential energy functions:

a. The ramped step potential[18]

$$V(x) = \begin{cases} 0 & \text{for} & x \le 0 \\ V_0 x/L & \text{for} & 0 \le x \le L \\ V_0 & \text{for} & L \le x. \end{cases} \tag{4.28}$$

b. The "quantum bouncer"[19]

$$V(x) = \begin{cases} \infty & \text{for } x \le -L \\ K(x + L) & \text{for } -L \le x. \end{cases} \tag{4.29}$$

c. The infinite square well with a thin barrier at its center.[20]

4.43 Other Sorts of Wave Packets

Use the "user defined wave function" feature to investigate initial wave packets that —

a. are not symmetric about $\langle x \rangle$. (For example a nonsymmetric tent.)

b. are of the form $(A(x) + iB(x))e^{ikx}$ with $A(x)$ and $B(x)$ real.

4.44 Display of Energy Uncertainties

Examine the code for finding position uncertainties in procedure **Find-MeanXP** and the code for displaying them near the end of procedure **PlotWFandPotl**. Invent and implement a method for displaying the uncertainty in energy.

4.45 Split Expectation Values

In some scattering simulations it is useful to know the expected position for the part of the wave function to the left of the scatterer and for that part to the right of the scatterer. Modify procedures **FindMeanXP** and **PlotWFandPotl** to implement this feature.

4.46 The Nonconservation of Energy

Set the potential to infinite square well (with default parameters) and the initial wave function to Gaussian (with default parameters). Set the display to show the energy spectrum (probabilities only), and run the simulation. The exact energy probabilities do not change with time, and the algorithm is impressive in echoing this behavior. Now change the width of the well to 0.80 nm, and try again. What went wrong? (Hint: Where are the well edges in relation to the grid points?)

4.47   Time-Dependent Potentials

Study the section 4.7 and the procedure **PrepareWithDriving**. Then modify **QMTime** to handle an expanding or contracting infinite square well.

4.48   Box Wave Function

Use the "user defined wave function" feature to produce the initial wave function

$$\Psi_0(x) = \begin{cases} 0 & \text{for} & x < 0 \\ He^{ikx} & \text{for} & 0 \le x \le L \\ 0 & \text{for} & L < x. \end{cases} \tag{4.30}$$

Show analytically that $\Delta x = L/2\sqrt{3}$ and that the expectation energy is infinite. How does this infinite energy show up numerically? (Hint: Vary the grid size.)

4.7   Appendix: Discretization of the Schrödinger Equation

4.7.1   A Naive Discretization

In our computer representation of a wave function, we store the value of

$$\Psi_j(t) \approx \Psi(x_j, t) \tag{4.31}$$

only for certain "grid points":

$$x_j = x_{\min} + (j-1)\Delta x, \quad j = 1, 2, 3, \dots, N. \tag{4.32}$$

How do these values change after a time Δt has passed? In other words, having already discretized space and time, how should we discretize

$$i\hbar \frac{\partial \Psi(x, t)}{\partial t} = -\frac{\hbar^2}{2m} \frac{\partial^2 \Psi(x, t)}{\partial x^2} + V(x, t)\Psi(x, t), \tag{4.33}$$

the Schrödinger equation?

It is clear that we should approximate the time derivative at point x_j by

$$\frac{\partial \Psi(x_j, t)}{\partial t} \approx \frac{\Psi_j(t + \Delta t) - \Psi_j(t)}{\Delta t}. \tag{4.34}$$

The spatial derivatives are only a bit more tricky. To approximate $\partial^2 \Psi / \partial x^2$ at point x_j, we begin by approximating $\partial \Psi / \partial x$ at $\frac{1}{2}\Delta x$ to the left of x_j and at $\frac{1}{2}\Delta x$ to the right of x_j:

$$\frac{\partial \Psi}{\partial x}(x_j - \tfrac{1}{2}\Delta x, t) \approx \frac{\Psi_j(t) - \Psi_{j-1}(t)}{\Delta x} \tag{4.35}$$

$$\frac{\partial \Psi}{\partial x}(x_j + \tfrac{1}{2}\Delta x, t) \approx \frac{\Psi_{j+1}(t) - \Psi_j(t)}{\Delta x}. \tag{4.36}$$

The approximation for $\partial^2 \Psi / \partial x^2$ is the difference between these values divided by Δx:

$$\frac{\partial^2 \Psi}{\partial x^2}(x_j, t) \approx \frac{\Psi_{j+1}(t) - 2\Psi_j(t) + \Psi_{j-1}(t)}{\Delta x^2}. \tag{4.37}$$

When Eqs. 4.32, 4.33, and 4.36 are put together, we obtain the "discrete Schrödinger equation"

$$i\hbar \frac{\Psi_j' - \Psi_j}{\Delta t} = -\frac{\hbar^2}{2m} \frac{\Psi_{j+1} - 2\Psi_j + \Psi_{j-1}}{\Delta x^2} + V(x_j, t)\Psi_j, \tag{4.38}$$

where we have used Ψ_j for $\Psi_j(t)$ and Ψ_j' for $\Psi_j(t + \Delta t)$.

This discretization of the Schrödinger equation has many attractive features: it is simple, direct, and easy to code into a program. Unfortunately, it is also useless. This discretization turns out to be *unstable*: as time evolves and the equation is iterated over and over, any tiny error present in the initial wave function will grow exponentially. Soon the error will swamp the solution. (And there is certain to be error in the initial wave function ... round off will see to that!) That mathematical cause of this problem is not hard to uncover[22] but it can be demonstrated even more easily:

4.49 Exercise: Instability of the Naive Discretization

Program **QMTime** comes with two procedures to step the wave function forward in time: **StepWF**, the usual (and stable) one, and **StepWFNaive**, which uses the scheme shown above (Eq. 4.37). Edit **QMTime**, changing all invocations of **StepWF** to **StepWFNaive**. Recompile the program, then run it to see what goes wrong. Is energy conserved? How about probability?

4.7.2 The Crank–Nicholson Discretization

Equation 4.37 is not the only way to discretize the Schrödinger equation. Many different *stable* discretization schemes[23–27] exist; the one used in **QMTime** is the "Crank-Nicholson" scheme.[17,21,22] In this technique, the right-hand side of Eq. 4.37, which is a function of time, is replaced by the average of that right-hand side evaluated at time t and at time $t + \Delta t$, to form

$$i\hbar \frac{\Psi_j' - \Psi_j}{\Delta t} = \frac{1}{2}\left[-\frac{\hbar^2}{2m\Delta x^2}(\Psi_{j+1} - 2\Psi_j + \Psi_{j-1} + \Psi_{j+1}' - 2\Psi_j' + \Psi_{j-1}') \right.$$

$$\left. + V(x_j, t)\Psi_j + V(x_j, t + \Delta t)\Psi_j' \right]. \tag{4.39}$$

It is clear intuitively that the Crank-Nicholson discretization (Eq. 4.39) ought to be more accurate than the naive discretization (Eq. 4.38). It is not clear, but is nevertheless true, that the Crank-Nicholson discretization is stable for any value of Δx or Δt. (Note that for large values of Δx or Δt Crank-Nicholson will be stable but not accurate: the difference between these terms is discussed in Press et al.,[17] section 1.2.)

In the naive technique it was obvious how to find the new variables Ψ'_j in terms of the old variables Ψ_j. It is not so easy in Crank-Nicholson: to do so we must solve a system of N linear algebraic equations for the N unknowns Ψ'_j. Rearrange Eq. 4.38 to form

$$\Psi'_{j+1} + D_j(t + \Delta t)\Psi'_j + \Psi'_{j-1} = R_j(t), \tag{4.40}$$

where

$$D_j(t) = -2 - \frac{2m\Delta x^2}{\hbar^2}V(x_j, t) + i\frac{4m\Delta x^2}{\hbar\Delta t} \tag{4.41}$$

(the name D refers to "diagonal matrix element") and where

$$R_j(t) = -\Psi_{j+1} - D_j^*(t)\Psi_j - \Psi_{j-1} \tag{4.42}$$

(the name R refers to "right-hand side"). The N equations of form 4.40 are particularly easy to solve because of their tridiagonal form.

(Program **QMTime** is written with one small deviation from the Crank-Nicholson algorithm as presented here. Instead of using $D_j(t)$ in Eq. 4.42 and $D_j(t + \Delta t)$ in Eq. 4.40, the program uses the average $(D_j(t) + D_j(t + \Delta t))/2$ in both places. This affects the results only if the potential energy depends on time.)

4.50   **Exercise: Edge Effects**
How is Eq. 4.40 modified if j is the first, second, last, or next-to-last point on the grid?

4.51   **Exercise: Solution of Tridiagonal Systems**
If D_j and R_j, $j = 1, 2, \ldots, N$, are known, and U_j and A_j are available work arrays, show that the Equation 4.40 can be solved through the following algorithm:

{Gaussian elimination: Convert tridiagonal system to equivalent system with ones on the diagonal, U_j on the upper diagonal, and A_j on the right-hand sides.}
$U_1 := 1/D_1$;
FOR $j := 2$ **TO** N **DO**
    $U_j := 1/(D_j - U_{j-1})$;

$A_1 := R_1 * U_1$;
FOR $j := 2$ **TO** N **DO**
    $A_j := (R_j - A_{j-1}) * U_j$;

{Backsubstitution.}
$\Psi_N := A_N$;
FOR $j := N - 1$ **DOWNTO** 1 **DO**
    $\Psi_j := A_j - U_j*\Psi_{j+1}$;

(Inspection of the code in procedure **StepWF** will show, however, that even further economies can be achieved. The program does not store the array of right-hand sides R_j; instead the appropriate values are calculated just before they are needed. There is no auxiliary array A_j; instead the value of A_j is stored in Ψ_{j-1}, the value of which is no longer needed.)

4.7.3 Properties of the Crank-Nicholson Discretization

The Crank-Nicholson updating scheme has two very desirable properties: it conserves probability and (when energy is suitably defined) it conserves energy. In this subsection we first cast Crank-Nicholson into a form that facilitates analytic use, and then use that form to prove the above-mentioned theorems. (Our proofs will apply only to cases in which the potential energy is time-independent.)

Equation 4.40 is a good way to represent the Crank-Nicholson technique in discussions of coding, but it is hard to use it to prove theorems. For such purposes Crank-Nicholson is best cast into matrix form. Denote the column matrix of wave function grid values by $\boldsymbol{\Psi}$: the jth element of $\boldsymbol{\Psi}$ is Ψ_j, the jth element of $\boldsymbol{\Psi}'$ is Ψ'_j. Now define a square Hamiltonian matrix $\mathbf{H}$ with elements

$$H_{j,j'}(t) = -\frac{\hbar^2}{2m\Delta x^2}(\delta_{j+1,j'} - 2\delta_{j,j'} + \delta_{j-1,j'}) + V(x_j, t)\delta_{j,j'}. \qquad (4.43)$$

In terms of this real symmetric matrix, the Crank-Nicholson discretization (Eq. 4.39) is

$$\frac{i\hbar}{\Delta t}(\boldsymbol{\Psi}' - \boldsymbol{\Psi}) = \frac{1}{2}[\mathbf{H}(t)\boldsymbol{\Psi} + \mathbf{H}(t + \Delta t)\boldsymbol{\Psi}'] \qquad (4.44)$$

or

$$\left[1 + i\frac{\Delta t}{2\hbar}\mathbf{H}(t + \Delta t)\right]\boldsymbol{\Psi}' = \left[1 - i\frac{\Delta t}{2\hbar}\mathbf{H}(t)\right]\boldsymbol{\Psi}. \qquad (4.45)$$

From now on we shall assume that the potential energy function, and hence the Hamiltonian matrix, is independent of time.

4.52 Exercise: Other Discretizations in Matrix Form

a. Write the naive (or "forward time center space") discretization in matrix form.

b. A third discretization, beyond Crank-Nicholson or forward time center space, is the "fully implicit" discretization in which the right-hand side of Eq. 4.37 is evaluated at time $t + \Delta t$ instead of time t. Write this discretization in matrix form.

We wish to prove that time development through Crank-Nicholson conserves probability. In other words we need to show (using component notation) that

$$\sum_{j=1}^{N} \Psi'_j{}^*\Psi'_j\Delta x = \sum_{j=1}^{N} \Psi_j^*\Psi_j\Delta x \qquad (4.46)$$

or equivalently (using matrix notation) that

$$\boldsymbol{\Psi}'^{\dagger}\boldsymbol{\Psi}'\Delta x = \boldsymbol{\Psi}^{\dagger}\boldsymbol{\Psi}\Delta x. \qquad (4.47)$$

To do so we first write

$$\boldsymbol{\Psi}' = \left(1 + i\frac{\Delta t}{2\hbar}\mathbf{H}\right)^{-1}\left(1 - i\frac{\Delta t}{2\hbar}\mathbf{H}\right)\boldsymbol{\Psi} \qquad (4.48)$$

and its hermitian adjoint,

$$\Psi'^{\dagger} = \Psi^{\dagger}\left(1 + i\frac{\Delta t}{2\hbar}H\right)\left(1 - i\frac{\Delta t}{2\hbar}H\right)^{-1}. \tag{4.49}$$

It is, however, easy to show that the matrices

$$\left(1 + i\frac{\Delta t}{2\hbar}H\right) \quad \text{and} \quad \left(1 - i\frac{\Delta t}{2\hbar}H\right) \tag{4.50}$$

commute, and that if two matrices commute, then their inverses do also. Hence

$$\Psi'^{\dagger}\Psi' = \Psi^{\dagger}\left(1 + i\frac{\Delta t}{2\hbar}H\right)\left(1 - i\frac{\Delta t}{2\hbar}H\right)^{-1}\left(1 + i\frac{\Delta t}{2\hbar}H\right)^{-1}\left(1 - i\frac{\Delta t}{2\hbar}H\right)\Psi$$

$$= \Psi^{\dagger}\Psi \tag{4.51}$$

and the theorem is proved.

4.53   Exercise: Conservation of Probability in Other Discretizations

Show that neither the forward time center space discretization nor the implicit discretization conserves probability in general.

4.54   Exercise: Conservation of Energy

a. Prove that time development through Crank-Nicholson conserves mean energy, i.e., that

$$\Psi'^{\dagger}H\Psi'\Delta x = \Psi^{\dagger}H\Psi\,\Delta x. \tag{4.52}$$

b. Show that neither the forward time center space discretization nor the fully implicit discretization conserves energy in general.

4.55   Exercise: Definition of Energy

The previous exercise assumed that the mean energy was

$$\langle E \rangle = \Psi^{\dagger}H\Psi\,\Delta x. \tag{4.53}$$

a. Show that the definition above is equivalent to

$$\langle E \rangle = \frac{\hbar^2}{m\,\Delta x}\sum_{j=1}^{N}[\Psi_j^*\Psi_j - \mathrm{Re}\{\Psi_j^*\Psi_{j-1}\}] + \Delta x\sum_{j=1}^{N}V(x_j)\Psi_j^*\Psi_j \tag{4.54}$$

or, for normalized wave functions,

$$\langle E \rangle = \frac{\hbar^2}{m\,\Delta x^2} - \frac{\hbar^2}{m\,\Delta x}\sum_{j=1}^{N}\mathrm{Re}\{\Psi_j^*\Psi_{j-1}\} + \Delta x\sum_{j=1}^{N}V(x_j)\Psi_j^*\Psi_j. \tag{4.55}$$

b. Show that this definition is an appropriate discretization of the continuum formula

$$\langle E \rangle = \int_{-\infty}^{+\infty}\Psi^*(x)\left[-\frac{\hbar^2}{2m}\frac{d^2}{dx^2} + V(x)\right]\Psi(x)dx. \tag{4.56}$$

c. Show that an equivalent formula for the continuum mean energy is

$$\langle E \rangle = \int_{-\infty}^{+\infty} \left[\frac{\hbar}{2m} \frac{d\Psi^*}{dx} \frac{d\Psi}{dx} + \Psi^*(x)V(x)\Psi(x) \right] dx \,. \qquad (4.57)$$

d. Discretize this formula, and code that discretization into program **QMTime** as a replacement for function **MeanE**. Show by example that this discretization of energy is *not* conserved.

References

1. Yeazell, J. A., Stroud, C. R. Jr. Observation of spatially localized atomic electron wave packets. Physical Review Letters. **60:**1494–1497, 1988.

2. Schrödinger, E. *Collected Papers on Wave Mechanics*. New York; Chelsea Publishing Company, 1978, pp. 102–104.

3. Liboff, R. L. *Introductory Quantum Mechanics*. Reading, MA: Addison-Wesley, 1980, section 7.9.

4. Merzbacher, E. *Quantum Mechanics*, 2nd ed. New York: Wiley, 1970, section 2.3.

5. Baym, G. *Lectures on Quantum Mechanics*. Reading, MA: Benjamin, 1969, pp. 62–65.

6. Cohen-Tannoudji, C., Diu, B., Laloë, F. *Quantum Mechanics*. New York: Wiley, 1977, pp. 79–85.

7. Styer, D. F. Using computers to build insight. In *Computing in Advanced Undergraduate Physics*, ed. D. M. Cook, pp. 201–203, Appleton, WI: Lawrence University, 1990.

8. Elmore, W. C., Heald, M. A. *Physics of Waves*. New York: McGraw-Hill, 1969.

9. Bramhall, M. H., Casper, B. M. Reflections on a wave packet approach to quantum mechanical barrier penetration. American Journal of Physics. **38:** 1136–1145, 1970.

10. Goldberg, A., Schey, H., Schwartz, J. One-dimensional scattering in configuration space and momentum space. American Journal of Physics. **36:**454–455, 1968.

    Good, R. H. Momentum space film loops. American Journal of Physics **40:**343–345, 1972.

11. French, A. P., Taylor, E. F. *An Introduction to Quantum Physics*. New York: Norton, 1978, pp. 408–413, 422–423.

12. Howard, S., Roy, S. K. Coherent states of a harmonic oscillator. American Journal of Physics. **55**:1109–1117, 1987.

13. Yan, C. C. Soliton like solutions of the Schrödinger equation for simple harmonic oscillator. American Journal of Physics. **62**:147–151, 1994.

14. Styer, D. F. The motion of wave packets through their expectation values and uncertainties. American Journal of Physics. **58**:742–744, 1990.

15. Feynman, R. P., Leighton, R. B., Sands, M. *The Feynman Lectures on Physics*, volume III. Reading, MA: Addison-Wesley, 1965.

16. Deutchman, P. A. Tunneling between two square wells—computer movie. American Journal of Physics. **39**:952–954, 1971.

    Johnson, E. A., Williams, H. T. Quantum solutions for a symmetric double square well. American Journal of Physics. **50**:239–243, 1982.

17. Press, W. H., Flannery, B. P., Teukolsky, S. A., Vetterling, W. T. *Numerical Recipes*. Cambridge: Cambridge University Press, 1986, section 17.2.

18. Bolemon, J. S., Haley, S. B. More time-dependent calculations for the Schrödinger equation. American Journal of Physics. **43**:270–271, 1975.

19. Langhoff, P. W. Schrödinger particle in a gravitational well. American Journal of Physics. **39**:954–957, 1971.

    Gibbs, R. L. The quantum bouncer. American Journal of Physics. **43**:25–28, 1975.

20. Segre, C. U., Sullivan, J. D. Bound-state wave packets. American Journal of Physics. **44**:729–732, 1976.

21. Goldberg, A., Schey, H., Schwartz, J. Computer-generated motion pictures of one-dimensional quantum-mechanical transmission and reflection phenomena. American Journal of Physics. **35**:177–186, 1967.

22. Koonin, S. E. *Computational Physics*. Reading, MA: Addison-Wesley, 1986, sections 7.1, 7.2, and 7.5.

23. De Raedt, H. Product formula algorithms for solving the time dependent Schrödinger equation. *Computer Physics Reports*. **7**:1–72, 1987.

24. Leforestier, C. et al. A comparison of different propagation schemes for the time dependent Schrödinger equation. Journal of Computational Physics. **94**:59–80, 1991.

25. Richardson, J. L. Visualizing quantum scattering on the CM-2 supercomputer. Computer Physics Communications. **63**:84–94, 1991.

26. Tal-Ezer, H., Kosloff, R., Cerjan, C. Low-order polynomial approximation of propagators for the time-dependent Schrödinger equation. Journal of Computational Physics. **100:**179–187, 1992.

27. Visscher, P. B. A fast explicit algorithm for the time-dependent Schrödinger equation. Computers in Physics. **5:**596–598, 1991.

5

Electron States in a One-Dimensional Lattice

Ian D. Johnston

5.1 Introduction

One of the most useful problems to which elementary quantum mechanics can be applied is that of an electron inside a solid. It should be possible to talk about the physics of this situation relatively straightforwardly by focusing on the potential which determines the interaction of the electrons with the atoms making up the solid lattice, and solving a stationary Schrödinger equation. In other words, treat it as a problem of finding energy eigenvalues.

The hope is that an understanding of the stationary states of the electrons should give a great deal of intuition about how they *move* inside the solid. That will, in turn, lead to a theory of conductivity—obviously a field with immense applicability. In the real world, a solid is three-dimensional, but the only Schrödinger equation which is easy to solve is one-dimensional. However, it has long been realized that a one-dimensional model can be expected to show at least some of the same qualitative behaviour as three. Furthermore, because electrons that move through the whole solid are of most interest, the wave functions describing the less energetic of them would typically have wavelengths comparable with the dimensions of the body—certainly much greater than the typical distance between atoms. Therefore, the overall behavior should not depend very strongly on the exact shape of the potential well presented by each atom. It is easiest to treat these wells as squares or some other simple shape.

Therefore, in this chapter, we study the eigenfunctions and eigenvalues of an electron in a regular array of potential wells having various simple shapes. We will go through the basic approximations that have been devised to handle what is a forbidding analytical problem, and then describe the computational approach which makes it tractable.

5.2 The Band Theory of Solids

5.2.1 The Free Electron Model

In the development of the modern theory of electrical conduction the first insights came in the early 1900s, well before the advent of quantum mechanics. It was argued that, as an electron moves throughout the whole lattice, the most significant features of the potential it sees are the side walls which keep it confined within the body. In some sense the atomic wells might be considered of secondary importance. So, as a first approximation, it seemed sensible to ignore the atomic wells altogether, and treat the electron as though it were moving inside an empty box. This was the so-called **free electron model**, and it proved surprisingly successful.

A detailed discussion of what can be predicted with this simple picture can be found in many standard textbooks, for example, Kittel[1] or Slater.[2] Even though the free electron theory of conduction predated quantum theory, it is easy to recast it in a quantum mechanical form. It says that, to first approximation, the wave functions look like the standing waves on a violin string. Furthermore, the energy levels are closely spaced, typically with small energy separations, of the order

$$\Delta E \sim \left(\frac{h^2}{mL^2}\right)n\,,\qquad(5.1)$$

where L is a typical length of the lattice and n is an integer from 1 to ∞.

For an ordinary sample, L might be some centimeters, and n would be expected to be roughly the number of atoms along the length of the lattice—of order 10^6. So the levels might be separated by energies like 10^{-30} J or 10^{-11} eV. This is much, much smaller than typical thermal energies, so for most purposes, they form essentially a continuum of states. The process of electrical conduction, in which electrons move in response to an external electric field, involves their being raised to higher energy levels by this field. Statistical considerations have to be invoked at this point, but the key feature is that, because the levels are so close together, these upward transitions between states take place very easily. And that is why a solid described by this picture is a conductor of electricity.

5.2.2 The Bloch Theorem

The main inadequacy of this simple picture is that not all solids are conductors. Some are insulators. Presumably it is the effect of the potential wells which makes the difference. Clearly it is necessary to take this into account.

Within a year of the introduction of quantum mechanics, in 1928, Bloch published the first steps to the solution of this problem. His was a purely formal result. He showed that, if the boundaries of the solid were ignored, and the potential taken to be perfectly periodic over the whole range of x from $-\infty$ to $+\infty$, then the particular solutions of the Schrödinger equation have the form of a traveling wave, multiplied by a function with the same periodicity as the lattice (**Bloch waves**), and general solutions could be constructed from these.

$$\psi_k(x) = u_k(x)e^{ikx}\,.\qquad(5.2)$$

A full discussion of the Bloch theorem and its ramifications can be found in older textbooks, for example, Peirls.[3] The most important feature, in this context, lies in the nature of the wave number k appearing in the equation. This wave number depends on the energy assumed for the electron, and must satisfy whatever equation results when Eq. 5.2 is substituted into the Schrödinger equation. But there is no guarantee that this will yield a real number value for k. Indeed Bloch's successors soon found, by working with cases in which the atomic potential wells could be considered to be very weak, that there were whole ranges of the energy for which k was imaginary, and a wave function containing an exponential term that would diverge at either $\pm\infty$ could not be considered physically acceptable. Thus there arose the picture of **bands** of energies for which almost continuously allowable solutions existed, and **gaps** where there were none.

5.2.3 The Tight Binding Approximation

At much the same time, a similar conclusion was reached from another direction. In a row of atoms, each with a very strongly bound electron, the energy levels would change as the distance between neighboring atoms is altered. If they are very far apart, the energy levels available to any single electron would be those of an individual atom: though each level will actually correspond to a number of different states equal to the number of atoms in the row (it is said to be **degenerate**). But if the atoms are closer together, the individual atomic wave functions overlap slightly, causing the energies of all the different states in each level to be slightly different from one another (the degeneracy is said to be "split"). So the originally discrete energy level spectrum evolves into a number of quasi-continuous bands with gaps between them. This is called the **tight binding approximation**, and a good discussion of it can be found in Davydov[4] or Kittel.[1]

5.2.4 The Kronig-Penney Model

The first analytical demonstration of the more general case came in 1930, when R. de L. Kronig and W. G. Penney published a solution of the Schrödinger equation for the potential shown below.

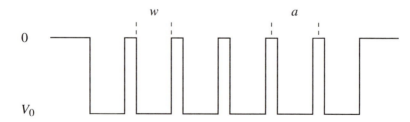

Their key assumption is that the solution is a Bloch wave. But the Bloch theorem really only applies to a strictly infinite lattice; therefore, the boundary condition that the wave function is zero at the edges of the crystal (or nearly so, as clearly it must be) cannot be imposed. Instead they made the simplifying assumption that the wave function repeats itself at the two ends. This is known as the assumption of **periodic boundary conditions**.

The beauty of the Bloch theorem is that the non-wavelike part of the solution (i.e., the function that multiplies the traveling wave) is periodic, so the Schrödinger equation only has to be solved in the region around one of the atoms. Therefore the problem resolves itself to a set of boundary matchings at the two sides of any one well, superimposed on the requirement that the wave function be periodic at the edges of the lattice. The argument can be followed in any of the standard textbooks, for example, Beard and Beard,[5] Hameka,[6] Kittel,[1] Merzbacher,[7] Sposito.[8]

The result of the calculation is usually quoted as the following transcendental equation, which must be satisfied by the wave number k and the energy E:

$$\cos(ka) = \cos(\lambda w) \cosh(\mu(a - w)) + \frac{\mu^2 - \lambda^2}{2\lambda\mu} \sin(\lambda w) \sinh(\mu(a - w)) \quad (5.3)$$

where

$$\lambda^2 \equiv 2m(|V_0| - |E|)/\hbar^2, \quad \text{and} \quad \mu^2 \equiv 2m|E|/\hbar^2.$$

The critical feature of this equation is that the left-hand side must lie between -1 and $+1$, but the right-hand side is not so limited. Therefore there are certain ranges of values of E for which no value of k can be found (i.e., no eigenfunction solution exists). Again it predicts energy bands and energy gaps. But the boundaries between the two can be worked out, by putting the left-hand side equal to ± 1 and solving for E.

5.2.5 Applications to Semiconductors

The most immediate use to which this theory was put was to the field of solid state physics, and in particular to the theory of electrical conduction (see chapters 4–6 of the solid state physics book in this series).

It is not the place to go into great detail here, since we are primarily interested in the quantum mechanical aspects. Suffice it to say that the concept of energy bands solved many of the obvious problems facing the early theories. The key idea lay in the Pauli principle, and the laws of quantum statistics (in particular that of Fermi and Dirac) which stemmed from it. This says that, by its very nature, only one electron can occupy any quantum state at any one time. Therefore the conduction properties of any solid would depend on the balance between the number of electrons capable of moving around inside the lattice, and the number of energy states they could occupy in any band. If there were more of the latter than the former, any electron could be raised in energy by the action of an external electric field, looking as though it were being "accelerated" and otherwise behaving like a classical charged particle in the presence of the field. The solid would be a conductor. But if there were not more states available than free electrons (the number of states is in fact determined by the number of atoms, and hence related to the number of electrons) then any one electron could not gain energy from any reasonable electric field because there would be no empty energy state it could go to. The substance would be an insulator.

In the 1940s and 1950s, interest shifted to a third kind of electrical behavior, that shown by **semiconductors**. Clearly, the ideas we have been developing were of importance here too. In some substances, the energy gaps may not be very wide, and therefore, in response to an applied field, some free electrons might have enough energy (simply from thermal fluctuations) to cross the gap. These substances would be very weak conductors, but their conductivity could be expected to increase with temperature (as opposed to ordinary conductors, whose conductivity decreases with temperature). This was just the behavior shown by a whole class of semiconductors, the so-called **intrinsic semiconductors**.

But even more interesting (from a theoretical point of view) were the so-called **extrinsic** (or impurity) semiconductors. In this class of materials it was found that the introduction of a very small number of atoms (as small a fraction as 1 part in 10^6) of a different atomic species into the lattice could change the conduction properties drastically. It was soon realized that extra energy states were being created by the impurity atoms which lay in the middle of the previously empty band gaps. These acted almost like a step ladder, allowing electrons to be "accelerated" across the gap more easily by "climbing" up these levels. Much theoretical attention was directed towards these new **intergap states**, and in particular to their *localization* properties.

It will be recalled that one very important feature about a Bloch wave is that it is "non-localized," i.e., it extends through the whole solid. This is important because it bears on the understanding of what happens in electrical conduction. An electric current should not be thought of as a number of classical particles or wave packets physically moving in one end of the lattice and out the other. A better model is a rearrangement of the electrons among stationary energy states, which give rise to effects classically asociated with motion, like magnetic fields, energy dissipation, and so on. But these new intergap states were clearly associated with particular sites in the lattice, and it was clear that the wave functions they gave rise to would also be strongly localized about those sites. Obviously this behavior had important consequences for an understanding of semiconductors. Starting with a classic paper by P. W. Anderson in 1958,[9] and going well into the 1970s, a number of very elegant analyses were published, of which an up-to-date discussion can be found in Mott.[10] In what follows, a model which uses the capabilities of modern personal computers to tackle these problems computationally will demonstrate this qualtiative behavior.

5.3   Computational Approach

5.3.1   Energy Levels in Regular Lattices

The computer program accompanying this chapter, **Latce1d**, uses the methods outlined in chapter 1 of this book to calculate the bound state wave functions and energy eigenvalues for an electron in a lattice consisting of a number of square potential wells evenly spaced in one dimension. It is important to remember, especially when comparing the results of these calculations with what is in ordinary textbooks, that the wave functions this method gives are all real.

In this program, the default well shape is simple square, and the default values of the width, separation, and depth of each well are fixed at 0.075 nm,

0.025 nm, and 300 eV, respectively. These values have no particular significance in themselves: they are simply convenient for demonstrating the qualitative features of the lattice behavior. However, the number of wells in the lattice may be changed, up to a maximum of 12 (beyond which the number of steps in the integration procedure must be increased, and the simplicity of the approach starts to be lost). This enables investigation of the tendency of energy levels to arrange themselves in bands, and to reproduce, without the mathematical assumptions that went into the Kronig-Penney approach, the qualitative behaviour described in section 5.2.5.

In these calculations, it is not intrinsically important that eigenvalues be calculated to a high degree of accuracy. All that is really needed is an approximate value for where the bands lie and the distribution of levels within them. However, the shape of the wave functions *is* important. It is found, particularly for large lattices, that some pairs of levels lie very close together and, unless the integration algorithm is sufficiently accurate, "mixing" of these pairs occurs, leading to wave functions that do not have the expected symmetries. (The term "mixing" is used here because it is the same kind of thing that occurs in perturbation theory calculations of degenerate, or nearly degenerate, states.) The "half-step" integration procedure described in Eq. 2.31 of chapter 2 is good enough for simple lattices, but the more accurate Numerov method was chosen here in order to handle more extreme conditions that users may like to try. Even so, in order for the program to run reasonably quickly, it was important to keep the number of integration steps as small as possible. It was found that about 400 steps is adequate, even for a lattice of 12 wells. But it must always be remembered that numerical results quoted below will be slightly different if any other number of integration steps or more accurate integrators are used.

5.3.2 Symmetries of the Wave Functions

The regular lattice as set up in this program, and the wave equation of which it forms a part, have a high degree of symmetry. It should be expected, then, that the solutions will also be very rich in symmetry properties. Indeed, that is why the solutions show Bloch-like standing wave structure. (Remember that the Bloch theorem itself was originally proved on what were essentially symmetry arguments.) When investigating the solutions, therefore, you should be on the lookout for the symmetries of the wave functions you find, particularly for lattices with small numbers of wells. In particular notice the properties of the wave functions of the lattice with only two wells, and compare them with the shapes of wave functions normally assumed for a simple molecule. To follow this further do Exercise 2 in this chapter.

5.4 Exercises

Many different calculations can be done with the program **Latce1d**. A few are listed here as suggested exercises. Some are actually projects that require modification of the program. Before trying the exercises, a review of the section on running the program will be useful.

5.1 **Single Square Well**

Start off with the default square well (width 0.075 nm and depth 300 eV) and the number of wells equal to 1.

    a. Use the **Method | Try Energy** menu item (i.e., select **Try Energy** from the pull-down menu **Method**), and find the energy eigenvalues yourself. You will observe the familiar energy level structure for a single square well.

    b. Use the automated binary search **Method | Hunt for Zero** to find individual states. Search between pairs of values spanning one of the eigenvalues you found.

    c. Use the menu item **Method | Solve Range of Energies**, choosing the range 0 to 300 eV in steps of 3 eV. Observe where the asymptote of the wave function crosses the axis, and compare those points with the values you just found. Confirm that there are three eigenvalues only, at binding energies of about 261, 151, and 8 eV. (Remember that the binding energy is the negative of actual energy, so the larger value corresponds to the ground state.) The eigenfunctions have the (also familiar) violin string standing wave shapes, except for a small tail at the edges.

5.2 **Two Square Wells**

Now choose two wells with the parameters defined above.

    a. There are five levels for this arrangement. Try to find them yourself, by the method described in the previous exercise.

    Hint: The lowest two levels are very close together, and you might have difficulty isolating them from one another. If so, use the menu item **Method | Solve Range of Energies**, this time choosing smaller energy steps.

    b. Now use the menu item **Method | Find Eigenvalues** in which you find the eigenfunctions automatically. You get the same five levels, two of which are very close to the previous ground state (at about 263 and 259 eV), two moderately close to the previous first excited state (158 and 140 eV), and one very near the top of the well (11 eV). (Note: To observe the actual values of these eigenvalues, you need to use the menu items **Spectrum | See Wave Functions** or **Spectrum | See Wfs and Probs**.) Why are the two lowest energies so close together, compared with the difference in energy between the two lowest states in the previous case?

    Hint: You should be able to see the answer by looking at the shapes of the two eigenfunctions. Notice that the wave functions themselves are quite dissimilar, but the probability functions are remarkably alike. This observation says that, so far as many measurable quantities are concerned, these two states should be very like one another.

c. Make each well narrower—set the width at 0.6 nm (**Wells | Well Parameters**). Again observe the complete spectrum. The two levels are now roughly 2 eV apart. Describe how this separation changes as the width of the wells decreases.

5.3   Lattice of Wells

Now choose 3 wells, and then 4, and then 5, and so on, up to 12, observing as you go through the sequence what happens to the energy level structure.

a. At each stage, if you have n wells, the eigenvalues arrange themselves in groups of n levels, clustered near the original values of 263 and 151 eV. Explain why this happens.

What you are seeing is obviously the existence of an energy band developing. From what you have observed so far, can you find a physical answer to the question, Why do energy bands and energy gaps exist?

b. Measure the band edge energies for lattices with a greater number of wells. How do they compare with the energy eigenvalues for the single well?

5.4   Eigenfunctions

Now go back through the same sequence of well numbers, this time concentrating on the shapes of the wave functions.

a. First of all you should notice that the states are always ordered with respect to their number of nodes: The ground state has no nodes, the first excited state 1, the next excited state 2, and so on. The lowest state in the second band always has a number of nodes equal to the number of wells in the lattice. Explain these observations.

b. Now compare the ground states for different values of n. They obviously have a common feature: They all look more-or-less like the fundamental violin string standing wave, multiplied by a function which repeats itself at each of the well sites. Similarly, look at all the lowest levels in the second band for different values of n. You should see the same thing, except that the repetitive function has a single node at each of the well sites. Observe what the lowest state in the third band looks like, and describe what you see.

c. Next describe what the second and third states in each band look like.

5.5   Bloch Waves

The pattern you should have noticed in the previous question was that each of the eigenfunctions, in so far as you could tell by just looking at them, was the product of—

- a standing wave whose wavelength decreases as you go up the band and repeats when you cross each band gap, and

- a periodic function, each repetitive element of which looks very like the eigenfunctions of a single well.

This clearly sounds very like the Bloch theorem. But it is not exactly the same. Strictly speaking, the Bloch theorem only applies to infinite lattices, and Bloch waves are traveling waves. The waves you found might be called **Bloch standing waves**.

Describe how you might construct the lowest-lying standing waves in each band from Bloch waves. Why can this construction never be exact but only an approximation?

5.6   The Kronig-Penney Model

Bloch waves were assumed by Kronig and Penney in their treatment of a lattice consisting of square wells. Clearly, that was a plausible assumption. Compare their analytic solution with what is calculated here. If you put the numerical values used in this program into Eq. 5.3, you should find that when the values of E make the right-hand side lie between ± 1, you can calculate the limits between which the corresponding energy bands should lie. How does your answer compare with the computed results?

5.7   Probability Densities

Choose the menu item **Spectrum | Sum Probabilities**, with which you can sum the *probability densities* for all levels within a specified range of quantum numbers.

a. Use this feature to sum the probability densities for all levels in lowest band. Repeat for all the states in the second band. In each case, the result is remarkably uniform across the lattice. Can you think what is the significance of this result? (Hint: Think about the charge density inside the material.)

b. From their definition Bloch waves were "non-localized"; they exist throughout the whole lattice. The eigenfunctions found here are not Bloch waves, and if you examine each one, they are not by any means uniform across the lattice. Do they still have the property of non-localization? What about the sum of probability densities calculated in the first part of this question?

5.8   Parabolic Wells

Change the shape of each well to parabolic (select **Wells | Shape of Each Well** and use the default values).

a. The wells now overlap one another, i.e., the range of each well is greater than the distance between "atoms." Explain why the graph of the total potential is different at the edge of the lattice from what it is at the center.

b. Does this difference show itself quantum mechanically? Is there, for example, any tendency for the wave functions to be more localized towards the center, when you sum the probabilities over any band?

(You should find that there is no such effect.)

c. Narrow the width at the top of each parabola so there is no overlap (make the width = 0.10 nm). Observe particularly the effect on the gap between the second and third bands.

d. Increase the width at the top of each parabola to 0.125 nm. Describe what happens to the band gap. What is the significance of this?

5.9 Coulombic Wells

Repeat the last exercise with coulombic wells.

a. Now there is a clear change of the bottom of the wells towards the edges of the lattice. Why does this occur?

(Hint: Remember that the Coulomb potential has a very long range and in this program is only truncated at the edge of the screen.)

b. Is there now any tendency for the states to be localized toward the edge?

You should find that there is a clear tendency for the higher levels in any band to be displaced toward the edges, even though, summed over the whole band, there is not. Why does this occur?

5.10 Density of States

Choose a lattice of 12 square wells.

a. Tabulate the energy differences between adjacent states. You should see that they tend to be closer together toward the top and bottom of the band.

b. Plot roughly the *density of states* (i.e., draw a crude histogram of the number of levels in a small energy range, as a function of energy). It is not easy to do this with only 12 states to work with, but try to make clear what you observed in part 1. You might try to divide the width of the band into five equal ranges and see how the 12 levels spread themselves among them.

c. There are theoretical expressions for the density of states which can be found in standard textbooks (e.g., Kittel[1]). How does the crude histogram compare?

5.11 Irregular Lattices

Use a lattice of 12 square wells and select the menu item **Lattice | Irregular**. This lets you investigate what happens when the regularity of the lattice is disrupted. Select **One Shallow Well**. One of the wells is made shallower than the others. It is a crude model of the lattice having been "doped" by the addition of an "impurity" atom.

Before you start, calculate the energy levels you would find if you worked with that single well by itself. Now find the complete eigenvalue spectrum for the whole lattice. Describe what has happened to the topmost energy level in each band. Why did it do this?

Explain what you think would have happened had the odd well been made deeper than the others? Run the program to check your intuition.

5.12   Intergap States

The behavior shown in the previous question is often described by saying that one of the states has "migrated" into the previously empty energy gap. In what sense is it accurate to say that it was the "top" level that moved, rather than one of the ones further down in the band?

What about the energy eigenvalues of the other states in the band. How accurate is it to say that they "didn't move out of the band"?

5.13   Localization of Intergap States

Now investigate the shapes of the wave function corresponding to the level that moved. It is clearly localized around the "impurity" site. Can you think of any physical reasoning that would have led you to expect this? (If you can't, don't forget that this was the very point which theoretical physicists took so long to untangle.)

Investigate the shapes of the wave functions remaining in the band. What would you say about any attempt to solve this problem analytically in the Kronig-Penney manner (assuming that the solutions could be approximately represented by Bloch waves)?

5.14   Band Properties in Impurity Lattices

Describe what happens when the probablilites over all the levels in each band are added together. Sum the probability densities (**Spectrum | Sum Probabilities**) for states 1 through 9. Then sum the probability densities for state 10 by itself. Repeat for states 11 through 19, and state 20.

What is the significance of this?

5.15   Sharpness of the Band Edge

For a single impurity in the lattice, measure the separation of the intergap state from the band, and describe how this changes as the strength of the impurity changes. Start off with the default impurity, i.e., one of the wells being 30 eV shallower than the others, and allow this energy difference to approach zero.

At what stage would you say that the top state has no longer "left the band"? The theoretical prediction is that this transition is very sharp. What do you observe?

5.16   Different Impurities

Investigate the effect of changing the properties of the "impurity" well.

a. Change the depth of the impurity well (**Lattice | Adjust Parameters**). Describe any changes to the band states and the intergap state.

b. Change the location of the impurity in the lattice. Again describe changes to the band states and to the intergap state.

c. Change the impurity well to one which is narrower than the others, and repeat these exercises.

5.17 Electric Fields

Apply an electric field to a regular lattice of 12 wells (**Lattice | Apply Electric Field**). A sensible value to choose is 10 V. (Note that, in order to keep the total potential within easily calculable bounds, a field strength which changes the potential by more than 20% is not allowed.) Describe the changes to the shape of the probability functions.

a. Comment on the "position" of states 1, 13, 25, as opposed to states 12, 24.

b. Textbooks on solid state physics describe the behavior of an electron in one of the states near the top of a band as having **negative effective mass**. Explain how this description is consistent with the "positions" of the states you just observed.

c. Is it possible to estimate the polarization as a function of field strength for individual states?

5.18 Amorphous Lattices

In many solids (for example, glass) the lattice is not a regular crystalline array. Instead it is *amorphous*. If you choose the menu item **Lattice | Irregular Lattice**. and select **Amorphous**, you get a lattice of equal wells distributed with randomly varying spacings between them.

a. For an amorphous lattice, plot the new density of states as you did in Exercise 5.10. Comment on any differences from the regular lattice.

b. Try "doping" this lattice as you did in Exercise 5.14. Do you notice any differences from what you observed with a regular lattice?

5.5 Details of the Program

5.5.1 Running the Program

On first loading the program you are presented with a credit screen which can be cleared by pressing any key or clicking the mouse.

The screen layout becomes similar to that shown in Figure 5.1 (though without the wave functions drawn). You are also presented with the following menu of choices:

File	Wells	Lattice	Method	Spectrum

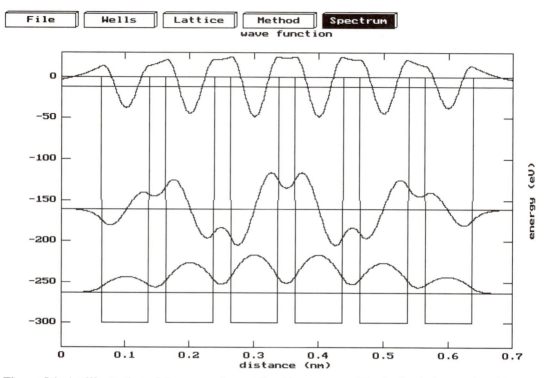

Figure 5.1: An illustration of the screen layout for this program. The lattice being explored is the default lattice of six square wells. Three eigenfunctions are shown: $n = 1$, corresponding to a binding energy of 263.51 eV; $n = 7$, to 160.86 eV; and $n = 13$, to 11.63 eV.

Move the highlighted item in the menu by the right and left <**arrow**> keys, and press **Return** when you have reached the one you want. Making the various choices has the following effects:

- **File**

  - **About CUPS**
    This gives a brief description of the CUPS project.

  - **About Program**
    This gives a brief description of what the program does.

  - **Configuration**
    This allows you to set a path for the storage of temporary files, to change the colors of the display, and to check how much memory has been used.

  - **Open...**
    This allows you to read in default values from a file you have previously saved from this program. Choose the name of the file from the list presented. If the file you want is not on the disk, press <**Esc**> to exit. The default values will not be changed.

  - **Save**
    This allows you to save all the values you have entered into all the input screens of the program, so that you can start where you left off when running

the program in future. If you opened a file earlier in the session, the information will be written into that file. If not you will be asked to supply a name for the file.

– **Save as...**
This allows you to save all the values you have entered into all the input screens of the program, so that you can start where you left off when running the program in the future. You will be asked to supply a name for the file.

– **Exit Program**
This takes you out of the program.

● **Wells**

– **Number of Wells**
This allows the user to choose how many wells will be in the lattice. The maximum number allowed is 12. Choosing **[Ok]** or pressing <**Enter**> will accept whatever number appears in the box on the screen, unless the number is greater than 12 or less than 1. If it is, the program will give an error message and wait for the number to be re-entered. Choosing **[Cancel]** or pressing <**Esc**> will abort entry and return to the main menu, with the number of wells unchanged from what it was before.

– **Shape of Each Well**
This allows the user to choose the shape of the wells making up the lattice. The choices are:

* **Square Well**

* **Parabolic Well**

* **Coulombic Well**

* **User Defined** (This item only appears if the user has set the **userFlag** within the program itself, see Section 5.5.2.)

– **Well Parameters**
Allows you to change the parameters of the well. Choosing **[Ok]** or pressing <**Enter**> will accept whatever numbers appears in the boxes on the screen, unless they are outside the allowed ranges (which vary from well to well). If it is, the program will give an error message and wait for the numbers to be re-entered. Choosing **[Cancel]** or pressing <**Esc**> will abort the entry and return to the main menu, with the well parameters unchanged from what they were before. Choosing **[View]** will display the lattice with the numbers just entered, but does not accept them permanently.

● **Lattice**

– **Regular Lattice**
This reverses any changes that have been made to the lattice, and constructs a regular lattice with the current number of wells, each of the current well shape and the default parameters appropriate to that shape.

– **Irregular Lattice**
This allows the user to choose a special lattice from the following list:

* **A Regular Lattice With One Shallow Well**

* **A Regular Lattice With One Narrow Well**

* **A Regular Lattice With One Well Displaced**

* **A Junction of Two Different Lattices**

* **An Amorphous Lattice**

* **User Defined** (This item only appears if the user has set the **userFlag** within the program itself, see section 5.5.2.)

– **Adjust Parameters**
This allows the two parameters associated with wells in the lattice, as well as their central position, to be adjusted individually. Numbers entered must be within the allowed ranges, and well centers must not be entered out of order. In either of these cases an error message will appear and the program will wait for new data entry.

– **Apply Electric Field**
This allows a small electric field (no larger than 20% of the maximum depth) to be applied across the whole lattice. Choosing **[Ok]** or pressing <**Enter**> will accept whatever number appears in the box on the screen, unless it is too large. If it is, the program will give an error message and wait for the number to be re-entered. Choosing **[Cancel]** or pressing <**Esc**> will abort entry and return to the main menu, with the lattice unchanged from what it was before. Choosing **[View]** displays the lattice with the number just entered, but does not accept it permanently.

• **Method**
This part of the program allows you actually to solve the Schrödinger equation for yourself. It is not strictly necessary if you are only interested in presenting physical results. It shows the methods by which those results were obtained and will be useful if you make changes to any of the program parameters. It allows four different submenu choices:

– **Try Energy (With Mouse)**
This allows you to select an energy by clicking within the graph. The Schrödinger equation is then solved and the solution drawn on the screen. The value of the energy chosen also appears in a message box at the top of the screen, together with the number of nodes. Three hot keys may be selected, with the mouse or by pressing—

* **Clear**   Redraw the screen without the previous solutions shown.

* **Zoom In/Out**   Increase/decrease the scale by a factor of two, useful when trying to home in on an eigenvalue.

* **Menu**   Return to the main menu.

– **Try Energy (From Keyboard)**
This allows you to enter an energy directly from the keyboard. The Schrödinger equation is then solved and the solution drawn on the screen. The value
of the energy chosen also appears in a message box at the top of the screen,
together with the number of nodes. Special results occur when the following
choices are made.

* **[Ok]**   Accept the number entered and show the solution, as just described.

* **[View]**   Redraw the screen without the previous solutions shown.

* **[Cancel]**   Return to the main menu.

– **Solve for Range of Energies**
This will solve the Schrödinger equation for energies within a given range,
so that you can see approximately where the eigenvalues are. A screen is presented asking you to supply three numbers—a lowest value for the binding
energy, a highest value, and a step size. (Positive numbers are required.) Default values of 0, 300, and 3 eV appear with the screen. Key in the values you
want and choose **[Ok]** or press <**Enter**>.

The program then integrates the Schrödinger equation using values of the energy within the range you entered, one solution for each "step size." It starts
with the boundary condition that y is zero at the left-hand side of the crystal
($x = 0$). When it has got to the right-hand side, it takes a note of the value
of y at this point (the asymptote) and plots it on the screen as a function of
energy. The places where the resulting curve crosses the energy axis (where
the asymptote is zero) indicate energy eigenvalues.

When it has solved the equation for all the appropriate energies in your range,
you may find the energy value for which this graph crosses the axis by pointing and clicking the mouse. You can't read off accurate values of these levels
from this curve, but you can get values close enough that you can use the next
part of the program effectively. When you have finished you can choose from
three hot keys:

* **Help**

* **Range**
Go back and repeat the plot with a different energy range or step size.

* **MENU**
Return to the main menu.

The program locates the points where the curve crosses the energy axis, redraws the potential curve, and marks on that diagram lines corresponding to
the energy levels it has found.

– **Hunt for Zero**
This allows you to find accurate eigenvalues by having the program hunt in
a binary search between two limits. A complete description of how it works
may be found in chapter 2, section 2.5.1.

● Spectrum

– **Find Eigenvalues**
The program automatically searches for the eigenvalues. It does this by a recursive procedure, so they appear on the screen in no fixed order. As each is found, the energy level is drawn on the graph and a beep is sounded. This sound may be disabled if it annoys (see **Sound** below).

– **See Wave Functions**
This item displays the eigenfunctions previously found for the particular lattice that has been chosen. On first entry it asks you which of the eigenfunctions you want to see, thus:

Choose a level number . . . . . .   [Clear] [Ok] [Cancel]

Enter from the keyboard. If you try to give it a number greater than the number of eigenvalues, or if you enter an alphabetic character, it will beep at you. When you have entered an acceptable number it will draw that eigenfunction, plot the probability, and display in the window the level number and the energy eigenvalue. It will also highlight on the potential plot the energy level in question. It will then wait for you to enter another level number. When you've seen enough eigenfunctions, choose **[Cancel]** or press <**Esc**> and you will return to the main menu.

– **See Wfs and Probs**
This does the same thing as the previous menu item except that it will plot the probability density as well as the wave function each time.

– **Sum Probabilities**
This item displays the the probability distributions of all the levels previously found for the particular lattice that has been chosen. On first entry it asks you for the range of levels you want to see. Enter from the keyboard in the usual way.

The screen layout changes at this point. The program first clears the top window and divides it into two. It shows all the levels on the lower half of the screen, and in the upper half, on top of one another, the probability distributions of all the energy states you chose. It marks on the potential plot which levels it is drawing by highlighting them in the drawing color. When it is finished it draws, on the same graph but in a different color, the sum of all these probabilities. A typical result is shown in Figure 5.2. (The important physical result is that, even though the individual states were all different, this sum is remarkably uniform across the lattice.) It then asks if you want to see the same thing done for any other group of states. When you've seen enough, choose **[Cancel]** or press <**Esc**> and you will return to the main menu.

– **Sound**
This toggles the sound on and off.

– **User Defined**
(The item appears only if the **userFlag** is set, see section 5.5.2.)

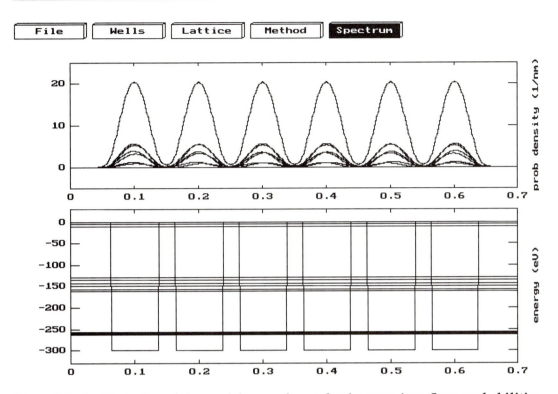

| File | Wells | Lattice | Method | **Spectrum** |

Figure 5.2: An illustration of the special screen layout for the menu item **Sum probabilities**. The probability densities corresponding to the lowest six eigenstates of the default lattice (levels shown in lower graph) are summed and the total displayed (upper graph).

5.5.2 Possible Modifications to the Program

As you become more used to working with this program, you may find the need to do calculations that the author did not think of. The program is designed to make it easy for you to go inside and add extra pieces of code of your own.

You may of course change any part of the code you choose, but there are a number of "user defined" procedures already written which you may use as templates. These are gathered together in the part of the program headed by the comment

******** USER DEFINED PROCEDURES ********

In that part of the code is a Boolean variable, **userFlag**. It is set to be *false*, but if you set it to *true*, at several points in the program an extra menu item will appear, called simply **User Defined**. By selecting that item you may access parts of the code which you yourself write.

There are three places where this occurs. In each case you are advised to write the specified procedure using the template that is there already, although you do not have to.

1. **Wells | Shape of Each Well**
   If you write a procedure called **CalculateUserDVector**, you can incorporate your own well. The template provided sets up a sinusoidal well, used in the Matthieu problem (see below).

2. **Lattice | Irregular Lattice**

   If you write a procedure called **UserLattice**, you can add your own particular combination of lattice parameters. The template provided makes one well deeper tham the others.

3. **Spectrum | User Defined**

   If you write a procedure called **UserOperation**, you can incorporate any operation you want to be carried out with the probability functions. The template provided calculates $<x>$ for any chosen eigenstate.

Note that, if you *really* want to customize the program to your own needs, you may also change the name "user defined" which appears in the menu items to something of your own choosing. Just follow the templates in the procedure **SetUserNames**.

Here are suggestions to show what you might do.

- Investigate a sinusoidal lattice (the Matthieu problem). This involves writing a procedure **CalculateUserPotential** to calculate a well whose shape is a sine wave with period equal to the spacing between atoms. If you want to compare the results with theory see Morse.[11] The code for this is already written as a template.

- Write a procedure **UserOperation** to use the calculation of probability functions to measure $\langle x \rangle$. This should allow you to calculate the polarization arising from an electric field, and eventually to estimate the effective mass.

References

1. Kittel, C. *Introduction to Solid State Physics*, 2nd ed. New York: John Wiley and Sons, 1961.

2. Slater, J. C. *Insulators, Semiconductors and Metals*. New York: McGraw-Hill, 1967.

3. Peirls, R. E. *Quantum Theory of Solids*. Oxford: Clarendon, 1955.

4. Davydov, A. S. *Quantum Mechanics*. Oxford: Pergamon, 1969.

5. Beard, D. B., Beard, G. B. *Quantum Mechanics, Principles and Applications*. Allyn & Bacon, 1970.

6. Hameka, H. F. *Quantum Mechanics*. New York: John Wiley and Sons, 1981.

7. Merzbacher, E. *Quantum Mechanics*, 2nd ed. New York: John Wiley & Sons, 1970.

8. Sposito, G. *Principles of Quantum Mechanics*. New York: John Wiley and Sons, 1970.

9. Anderson, P.W. Absence of Diffusion in Certain Random Lattices. Physical Review **109:**1492, 1958.

10. Mott, N. F. *Conduction in Non-Crystalline Materials.* Oxford: Clarendon, 1987.

11. Morse, P. M. The Quantum Mechanics of Electrons in Crystals. Physical Review. **35:**1310, 1930.

6

Three-Dimensional Bound States

Ian D. Johnston

6.1 Introduction

The behavior of a particle moving in a three-dimensional potential well can be studied by exactly the same methods as were used for the one-dimensional wells of chapter 2. Again we concentrate on situations which correspond to states of definite energy. The stationary Schrödinger equation is now a three-dimensional, second-order partial differential equation, the general solution of which requires considerably more work than in the one-dimensional case.

In chapter 2, the approach to the problem was abstract, in the sense that we described the behavior of a particle in a potential well, but never paid much atention to how such a potential might arise in practice. In the real world, however, one very easy interaction to envisage is that which occurs between two particles. Now the Schrödinger equation for two particles should, in principle, depend on six spatial coordinates, but this particular system is not difficult to handle. When there are no other forces except the mutual interaction between the two particles, the potential can only depend on the (vector) separation of the particles and not on their position vectors individually. This simplifies the mathematical problem enormously. As you can follow in any of the standard textbooks (e.g., Schiff,[1] pp. 88–90), the Schrödinger equation can be separated into two independent wave equations. One involves motion of the center of mass only, and the other describes the motion of a *single* object in a frame of reference in which the center of mass is at rest, whose mass is the **reduced mass** of the two particles. (There is an exactly corresponding result in classical mechanics, of course.)

The stationary Schrödinger equation for this object is then

$$-\frac{\hbar^2}{2\mu}\nabla^2\psi(\mathbf{r}) + V(\mathbf{r})\psi(\mathbf{r}) = E\psi(\mathbf{r}),\qquad(6.1)$$

where the reduced mass μ and the relative displacement $\mathbf{r}$ are defined in terms of the masses and position vectors of the two particles thus:

$$\mu \equiv \frac{m_1 m_2}{m_1 + m_2} \tag{6.2}$$

$$\mathbf{r} \equiv \mathbf{r}_1 - \mathbf{r}_2.$$

This is the equation that this program solves.

6.1.1 Separation of Variables

In most physical systems involving the interaction of two particles, there is a further mathematical simplification which can be made. Unless either particle has some property which is correlated with direction in the outside world (as it would have if, for example, it were spinning), then the interparticle potential can only depend on the *magnitude* of the relative displacement and not on its direction. In this case, when the operator ∇^2 is written out in spherical polar coordinates, the stationary Schrödinger equation becomes

$$\left[-\frac{\hbar^2}{2\mu} \frac{1}{r^2} \frac{\partial}{\partial r} \left(r^2 \frac{\partial}{\partial r} \right) + V(r) \right] \psi(r, \theta, \phi) + \frac{1}{2\mu r^2} \hat{\mathbf{L}}^2(\theta, \phi) \psi(r, \theta, \phi) = E\psi(r, \theta, \phi),$$
$$\tag{6.3}$$

where

$$\hat{\mathbf{L}}^2(\theta, \phi) \equiv -\hbar^2 \left[\frac{1}{\sin\theta} \frac{\partial^2}{\partial\phi^2} + \frac{1}{\sin^2\theta} \frac{\partial}{\partial\theta} \left(\sin\theta \frac{\partial}{\partial\theta} \right) \right]. \tag{6.4}$$

Since the quantities inside the first (square) parentheses of Eq. 6.3 depend only on r, and the operator $\hat{\mathbf{L}}^2$ depends only on θ and ϕ, the variables can be **separated** by the usual methods of partial differential equations (see, for example, Merzbacher[2]). This means that particular solutions of Eq. 6.3 can be written

$$\psi(r, \theta, \phi) \equiv R(r) Y(\theta, \phi), \tag{6.5}$$

where

$$\left[-\frac{\hbar^2}{2\mu} \frac{1}{r^2} \frac{\partial}{\partial r} \left(r^2 \frac{\partial}{\partial r} \right) + \frac{\lambda \hbar^2}{2\mu r^2} + V(r) \right] R(r) = ER(r) \tag{6.6}$$

and

$$\hat{\mathbf{L}}^2(\theta, \phi) Y(\theta, \phi) = \lambda \hbar^2 Y(\theta, \phi). \tag{6.7}$$

It is a fundamental result of the theory of differential equations that the most general solution of Eq. 6.3 can be expressed as a linear combination of solutions like Eq. 6.5. Therefore we need to discuss, and the program needs to compute, only the solutions of Eq. 6.6 and Eq. 6.7 independently.

6.1.2 Spherical Harmonics

The solution of Eq. 6.7 does not depend on the potential at all and is discussed extensively in all the standard textbooks (see, for example, Merzbacher,[2] pp. 178–189), and there is no need to repeat it here. The outcome of that discussion is that the eigenfunctions must be enumerated in terms of *two* quantum numbers *l* and *m*, and are written as

$$Y_l^m(\theta, \phi) = A_{lm}e^{im\phi}P_l^m(\cos\theta),\tag{6.8}$$

where the functions P_l^m are known as the **associated Legendre functions** and A_{lm} is a normalization constant, which depends on both *m* and *l*. The functions (Eq. 6.8) are called **spherical harmonics**. These functions are eigenfunctions, not only of the operator $\hat{\mathbf{L}}^2$, but also of $\partial/\partial\phi$:

$$\frac{\partial}{\partial\phi}Y_l^m(\theta, \phi) = imY_l^m(\theta, \phi)\tag{6.9}$$

$$\hat{\mathbf{L}}^2(\theta, \phi)Y_l^m(\theta, \phi) = l(l + 1)\hbar^2 Y_l^m(\theta, \phi).\tag{6.10}$$

If you compare Eq. 6.10 with Eq. 6.7 you can see that the quantity λ in Eq. 6.7 is related to the **rotational kinetic energy** of the system—i.e., the kinetic energy associated with purely tangential motion. Since rotational kinetic energy is equivalent to moment of inertia divided by angular momentum squared, we can say that the angular momentum of this system is given by

$$(\text{angular momentum})^2 = \lambda\hbar^2 = l(l + 1)\hbar^2.\tag{6.11}$$

Therefore the quantity *l* is known as the **orbital angular momentum quantum number**. The operator $\partial/\partial\phi$ can be shown to be associated with one component of the (vector) angular momentum and is closely correlated with the magnetic moment of the system. Therefore *m* is usually known as the **magnetic quantum number**.

There are important restrictions of the values these two quantum numbers can assume.

$$l = 0, 1, 2, 3 \ldots\tag{6.12}$$

$$m = 0, \pm 1, \pm 2, \pm 3 \ldots \pm l.\tag{6.13}$$

For the purposes of this chapter, we are not really interested in the mathematical properties of the spherical harmonics, merely their *shape*. The program can be asked to plot any of these functions, but their calculation does not change as different potential wells are chosen.

6.1.3   The Radial Wave Equation

The equation for the radial part for the wave function 6.6, can be greatly simplified by the following substitution:

$$u(r) \equiv rR(r),\tag{6.14}$$

whereupon that equation becomes

$$-\frac{\hbar^2}{2m}\frac{d^2u(r)}{dx^2} + \left[\frac{l(l + 1)\hbar^2}{2mr^2} + V(r)\right]u(r) = Eu(r).\tag{6.15}$$

This has exactly the same form as a one-dimensional Schrödinger equation for a particle moving in a combined potential, made up of the two-particle interaction $V(r)$ and the term $l(l + 1)\hbar^2/2mr^2$. This latter is sometimes called the

centrifugal potential, since the negative of its gradient is equal to the centrifugal force that a classical particle would "feel" when it is moving in a circular orbit with angular momentum $\hbar\sqrt{l(l + 1)}$.

There is one important difference between Eq. 6.15 and a one-dimensional Schrödinger equation, however. The radial coordinate r cannot have negative values. Hence the point $r = 0$ is a boundary of the equation. Furthermore, since $\psi(r, \theta, \phi)$ must be finite everywhere,* Eq. 6.14 says that $u(r)$ must vanish at $r = 0$. The only acceptable solutions of Eq. 6.15 are therefore those which satisfy this requirement.

For all numerical computations we will do, we will choose potentials $V(r)$ which are finite everywhere. This will involve a slight approximation when we come to work with the Coulomb potential, and all such calculations need to be done with an awareness of what error that approximation introduces. However, you will already have noticed that the centrifugal potential has a singularity at the origin, for all values of the quantum number $l > 0$. All radial wave functions, therefore, right at the origin, must look like the solution of

$$\frac{d^2u}{dx^2} - \frac{l(l + 1)}{r^2}u = 0. \qquad (6.16)$$

Attempting to solve this equation with a power series shows that the leading non-singular term for small r is

$$u(r) \approx Ar^{l+1}. \qquad (6.17)$$

Our numerical solutions of the radial wave equation will always take this as the boundary condition at $r = 0$ (See Fig. 6.1.)

Note that it would be expected that the eigenfunctions of Eq. 6.15 will be denumerated by a quantum number n, but the equation itself contains the orbital quantum number l. Therefore, the full set of radial eigenfunctions will depend on *two* quantum numbers, and will be written as $u_{nl}(r)$.

6.1.4 The Probability Cloud

A standard way to visualize these three-dimensional eigenfunctions is to plot the probability density of finding the particle anywhere on the $\phi = 0$ plane (i.e., the x-z plane). Because of the form of Eq. 6.8, this probability distribution is the same for all values of ϕ. That is, the probability of finding the particle between r and $r + dr$, θ and $\theta + d\theta$, ϕ and $\phi + d\phi$ is given by,

$$P(r,\theta,\phi)r^2 \sin\theta\, dr\, d\theta\, d\phi = |u_{nl}(r)|^2 |A_{lm}P_l^m(\cos\theta)|^2 r^2 \sin\theta\, dr\, d\theta\, d\phi. \quad (6.18)$$

It is convenient to plot this as a **scatter diagram**. This is done by placing a large number of points on the plane, whose positions are chosen randomly within the probability distribution given by Eq. 6.18. The resulting picture is usually interpreted as respresenting the "probability cloud" of the particle around the center of the potential, (see Fig. 6.2).

---

*This is not exactly true. However, the wave function must be square integrable, as has been pointed out in chapter 2, and at worst it can diverge only logarithmically at the origin.

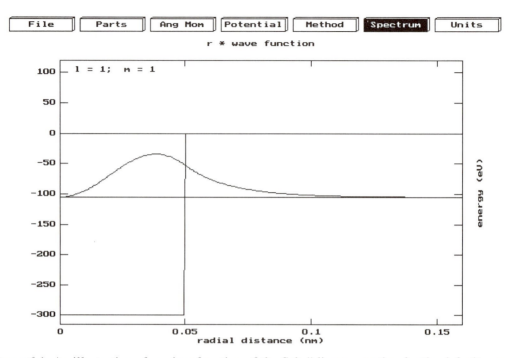

Figure 6.1: An illustration of an eigenfunction of the Schrödinger equation for the default square potential well. This solution corresponds to $l = 1$, $m = 1$ and a binding energy of 105.06 eV.

Note carefully what is being plotted in Fig. 6.2. We plot the probability of finding the particle at any point on the x-z plane. What you see is a section through the center of the particle, as it were. An alternative kind of diagram would have been to plot a *projection* of the probability onto the x-z plane, as though looking through a cloud of smoke. As an example of the difference, if we had a sphere of uniform probability throughout, the former method would produce a uniform disk, while the latter would produce one which is more intense at the center. It is the former method which is employed here.

6.1.5   Overlap Integrals

As explained in chapter 2, when quantum mechanics is *used*, much of the work which has to be done involves the calculation of overlap integrals of the form

$$\int_{all\ space} \psi^*(r,\theta,\phi)\hat{A}\psi(r,\theta,\phi)r^2 \sin\theta\, dr\, d\theta\, d\phi\,, \qquad (6.19)$$

where $\hat{A}$ is one of a number of different operators. These quantities form a **matrix** whose diagonal elements can be used to predict the possible results of measurement, and whose off-diagonal elements help in, for example, calculating the probability of transition from one state to another.

The program allows the calculation of different overlap integrals for those operators which depend on r only and not on θ and ϕ. For such operators, the known

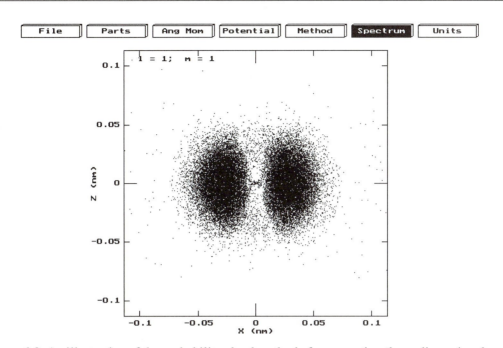

Figure 6.2: An illustration of the probability cloud method of representing three-dimensional wave functions. The wave function plotted is that illustrated in Figure 6.1. It has $l = 1$ and $m = 1$, and appears as a scatter diagram in the x-z plane. The complete, three-dimensional probability distribution is obtained by mentally rotating the diagram around the z-axis.

orthonormality of the spherical harmonics can be assumed, and all overlap integrals simplify to the following form:

$$\int_0^\infty u_{n_1 l}^*(r)\hat{A}u_{n_2 l}(r)\,dr. \tag{6.20}$$

These are the overlap integrals which are calculated by the program **Bound3d**.

6.2 Computational Approach

6.2.1 Numerical Solution of the Radial Equation

Because Eq. 6.15 is so similar to a one-dimensional equation, the same numerical techniques are used here as were described in chapter 2, with one difference. A wider range of *physical* situations are covered by this program, and the integration procedure will be called upon to work properly under a great variety of conditions. In particular it may happen that a very narrow well will be chosen, such that the integration procedure is asked to integrate the Schrödinger equation over a long range of r where the solution may be an increasing exponential. Under these conditions, many integration procedures will tend not to follow the other solution, which is a *decreasing* exponential. Therefore the method of solution used here involves integrating the equation from $r = 0$ *and* from $r = \infty$, and matching the two, in the usual way, at some point in between.

This algorithm is used at every point where a known eigenfunction is being calculated—where it is known that the solution for large r must be a decreasing exponential. When the program is *finding* eigenvalues, either under the user's control or automatically, this integration algorithm may be cumbersome to apply. Therefore the method used in chapter 2 is employed—of integrating to the right hand side of the screen and then trying to match to a decreasing exponential. It is best to be aware of this because if you pick a particularly difficult potential shape, the numerical values of the eigenvalues found by the two parts of the program may be very slightly different from one another.

6.2.2   Units and Scales

Since the program handles the three-dimensional Schrödinger equation, it is possible to deal with a wider range of physical problems than in one dimension. In particular the user may want to describe the behavior of an electron bound in an atom (or some structure the size of an atom), an atom bound inside a molecule, or a nucleon bound inside a nucleus.

In order to cover all these cases users are able to choose whichever units they want to work with:

- **atomic units**, where masses are measured in multiples of the electron mass, lengths are measured in nanometers, and energies in eV;

- **molecular units**, where masses are measured in multiples of the atomic mass unit, lengths are measured in nanometers, and energies in eV; or

- **nuclear units**, where masses are measured in multiples of the atomic mass unit, lengths are measured in femtometers, and energies in MeV.

Users also have the ability to choose the scale along the r-axis they wish to use in the integration. It is necessary to be able to change this feature because some potentials (like the Coulomb potential) have a long **range**—i.e., they fall off so slowly with distance that their eigenfunctions extend a long way from the center of the potential. Others (like most nuclear potentials) have only a very short range, and their eigenfunctions are confined to a narrow region near the origin. You will get best results from the program if you choose a range along the r-axis appropriate to the potential you are using.

6.3   *Exercises*

Many different calculations can be done with the program **Bound3d**. A few are listed here as suggested exercises. Some are actually projects that require modification of the program. Before trying the exercises, a review of the section on running the program will be useful.

6.3.1 Finding Eigenvalues and Eigenfunctions

6.1 Square Well

Work with the default square well, whose parameters are depth 300 eV, radius 0.05 nm. However, before you start, work out the energy levels of an *infinite* well of the same radius.

a. Demonstrate what solutions of the wave equation look like for different energies (use the menu item **Method | Try Energy**). Try to home in on one eigenfunction.

b. Next do an automated binary search (**Method | Hunt for Zero**), and find the energy levels of the well. Note:

$$\text{with } R_0 = 0.05 \text{ nm}, \quad E = \frac{h^2}{8m_e R_0^2} \approx 144 \text{ eV}.$$

Search around this value above bottom of well. Is there likely to be more than one eigenvalue for this well? Explain your answer.

c. Find the complete eigenvalue spectrum (**Spectrum | Find Eigenvalues**). Examine and describe the eigenfunction(s) (**Spectrum | See Wave Functions**).

d. Can you explain why the solution is so much lower in energy (i.e., of greater binding energy) than the infinite well of the same width?

6.2 Probability Cloud

For the square well you just investigated, carry out the following computations.

a. Use (**Spectrum | Probability Cloud**) to plot the probability cloud. Explain what you see.

b. Now use the menu item (**Ang Mom | Set Orbital Ang Mom, l**) to change the orbital angular momentum quantum numbers (l) to 1. Before running the program, sketch the equivalent one-dimensional potential well by adding together the potential $V(r)$ and the effective centrifugal potential (see Eq. 6.15).

Now find the eigenvalues. Why is the energy level for $l = 1$ of higher energy (lower bunding energy) than for $l = 0$?

c. Plot the probability cloud again, and explain the shape you see. Compare this shape with the three-dimensional plot of the spherical harmonic (menu item **Ang Mom | Display Spherical Harmonic**).

d. Change the magnetic quantum number (m) to ± 1 and do the same thing. This will not require you to recalculate the eigenvalues, because, although the Schrödinger equation depends on *l*, it doesn't depend on *m*.

e. Change *l* to 2. Again check the equivalent one-dimensional potential.

When you try to find the eigenvalues, nothing happens. Why is no energy level for *l* = 2?

6.3 Parabolic Well

Choose a parabolic well from the program menu (**Potential | Parabolic**). Choose parameters for this well (**Potential | Vary Well Parameters**) of depth = 600 eV and radius at the top = 0.15 nm.

a. Demonstrate first that the levels are equally spaced, and then that there is close (but not perfect) agreement with theoretical eigenvalues. This occurs because the theoretical result assumes the potential keeps increasing toward *r* = ∞, whereas the computational potential is truncated and set to zero outside the range.

b. Find one of the eigenfunctions using the **Method | Hunt for Zero** option. Measure positions of nodes of the different eigenfunctions by choosing the menu item **Method | Examine Solution**. Compare the results with the zeroes of the corresponding Hermite polynomials.

c. Find all the eigenvalues automatically **Spectrum | Find Eigenvalues**. How do these compare with theoretical values?

d. Now change the orbital angular momentum quantum number (*l*) to 1. What does theory predict the eigenvalues will be now? Compute the eigenvalues and check that they agree with the theoretical result.

6.4 Coulombic Well

Set up a Coulombic potential (**Potential | Coulombic**). Be very clear in your mind before you start that the potential the program constructs is not a "real" Coulombic potential. It is not allowed to become infinite at *r* = 0 (it is set, arbitrarily, to become constant when the formula would take it below −600 eV). And it is truncated at the far right of the screen, being set to zero at all points beyond. When you first select this potential the "edge of the screen" is at 0.32 nm.

a. Verify that the default parameters are appropriate for a simple hydrogen atom (i.e., one electron moving in the potential of a single proton).

b. When you find "all" the eigenvalues using the program as it is set up, you only get two. Why? *This is a most important concept to understand. The Coulomb potential is different from most other potentials that you will have to deal with because it has an extremely long* **range**.

Indeed, a careful definition of what that term means reveals that the range of this potential is infinite; there is no distance beyond which it can be ignored. For present purposes it means that successive bound states spread out further and further from the origin. The expectation value of the radius can be shown to depend as n^2.

When we calculate states of this potential numerically we adopt the convention of truncating the potential at the edge of the screen. Therefore we will always calculate no more than a few of the "real" states of this potential, no matter where we truncate it.

c. Check how closely these two values calculated so far agree with the theoretical values.

d. Using the menu item (**Units** | **r Axis**), change the maximum value on the r-axis to 1.00 nm. Recalculate the eigenvalue spectrum. How many states do you find now? How accurate are the values that you compute?

Note that you cannot keep increasing the r-axis too far else the procedure which draws tic marks along the axis will malfunction. You should, however, appreciate that the larger you make the maximum value of the r-axis, the less accurate the calculation of the closer-in state energies will be (because of the long range over which the method of solution must remain stable against following the solution which increases with increasing r—see the discussion in section 6.2.1).

e. Return to the lower states, change the orbital angular momentum quantum number, and recalculate the eigenvalues. You should observe a most important symmetry for the energy eigenvalues, viz., that the lowest energy with $l = 1$ is exactly equal to the second lowest energy for $l = 0$ (and similar results). This symmetry is a special feature of the Coulomb potential. Where does it come from?

6.5 Lennard-Jones Well

The Lennard-Jones potential, defined by

$$V(r) = -V_0 \left(\left(\frac{a}{r} \right)^6 - \left(\frac{a}{r} \right)^{12} \right),$$

is an approximation to the shape of the potential which binds two atoms into a molecule. (Refer, for example, to Brehm and Mullin,[3] p. 520, for a description of this potential.) In the program, the code which sets up this potential uses as default units, distances in nm, energies in eV, and masses in multiples of the a.m.u. (atomic mass unit).

a. Set up this potential (**Potential** | **Lennard-Jones**) using the default parameters. Notice that, whereas the theoretical potential is infinitely repulsive at small values of r, the computational one is truncated. Do you expect this to have a significant effect on the eigenvalues? Why?

b. Compute the eigenvalue spectrum. From the values that result, predict what frequency you would observe if *transitions* occurred between any two levels. Where are these frequencies in the electromagnetic spectrum?

c. Compute the same frequencies if the molecule were composed of two oxygen atoms bound together.

d. Change the orbital angular momentum of the molecule and calculate the energy eigenvalues again. What is the difference?

You should have noticed that the value of the energy levels did not change very much. When you investigated the square well in Exercise 2, changing *l* made a big difference to the binding energy. Why is the effect so much less in this case? *Hint:* Draw the equivalent one-dimensional potential.

6.6 Dumbell Well

The well which results from choosing **Potential | Dumbell** is a truncated parabolic well, with parameters chosen so that it is reasonably close to the Lennard-Jones well near the value of *r* for which the potential is minimum.

a. Construct an argument, based on the behavior you observed in the previous exercise, to say that you would expect the low-lying eigenvalues to be very similar.

b. Test this conclusion by calculating the eigenvalues for two "oxygen" atoms bound by this potential.

c. Explain why spectral lines emitted by molecules in this frequency range are usually referred to as the **vibrational spectrum**.

d. Change the orbital angular momentum of the "molecule" and calculate the eigenvalues again. What difference do you notice?

e. From the observations you made in this exercise and the previous one, predict approximately what you would expect the *rotational spectrum*—i.e., the spectral lines arising from transitions between states with different *l* values, but other quantum numbers the same— would be. Explain this on theoretical grounds.

6.7 Yukawa Well

The Yukawa potential, defined by

$$V(r) = -\frac{V_0}{r}e^{-r/a},$$

is a theoretical potential for the force arising from the strong interaction between two nuclear particles. (Refer, for example, to Eisberg and Resnick,[4] pp. 685–692, for a description of this potential.) In the program, the code which sets up this potential uses as default units, distances in fm, energies in MeV, and masses in multiples of the a.m.u.

    a. Set up this potential (**Potential | Yukawa**) using the default parameters. Calculate the eigenvalue spectrum.

    b. The *deuteron*, which consists of a proton and a neutron bound together, is known to have only one eigenvalue at a binding energy of 2.226 MeV. Various experiments suggest that the range of the force that binds them is about 1 fm. Calculate the depth of a Yukawa well that will give this result. (Use this range for the potential and choose a depth that will give one bound state at the required binding energy. Don't forget to use the reduced mass in your calculation.)

6.3.2 Properties of Bound State Functions

6.8 Orthonormality

Run part 2 of the program dealing with wave function properties, using the **parabolic well** with depth 300 eV and radius at the top 0.15 nm.

    a. Choosing the menu items (**Psi 1 | Eigenstate,n=1** and **Psi 2 | Eigenstate,n=1**), select the ground state energy eigenvalue for both input variables. Verify that the program has correctly normalized the ground state eigenfunction. Similarly, verify that the other two eigenfunctions are correctly normalized.

    b. Now verify that all three eigenfunctions are orthogonal to one another. (Since there are three different eigenfunctions, there are three different overlap integrals that you must look at.) How accurately do these integrals vanish?

6.9 Expectation Values

Choose a Coulombic potential that mimics the hydrogen atom.

    a. Measure the expectation value of r in all of its energy eigenstates, with $l = 0$. Compare these values with theoretical values.

    b. Now choose $l = 1, 2, \ldots$. Again measure $<r>$, and compare with theoretical values.

6.4 Details of the Program

6.4.1 Running the Program

On first loading the program you are presented with a credit screen which can be cleared by pressing any key or clicking the mouse. You are presented with the following menu of choices:

    File   Ang Mom   Potential   Method   Spectrum   Units

Choose one of these items either with the mouse, or by moving the highlighted item in the menu by the right and left <**arrow**> keys, and pressing <**Enter**> when you have reached the one you want. There are two different parts of the program, and this main menu changes depending on which part of the program you are currently working in. The first two menu items are common to both parts. Choosing either of them has the following effects:

● **File**

  – **About CUPS**
  This gives a brief description of the CUPS project.

  – **About Program**
  This gives a brief description of what the program does.

  – **Configuration**
  This allows you to set a path for the storage of temporary files, to change the colors of the display, and to check how much memory has been used.

  – **Open...**
  This allows you to read in default values from a file you have previously saved from this program. Enter the name of the file when requested. If the file you name is not on the disk, the default values will not be changed.

  – **Save**
  This allows you to save all the values you have entered into all the input screens of the program, so that you can start where you left off when running the program in future. If you opened a file earlier in the session, the information will be written into that file. If not you will be asked to supply a name for the file.

  – **Save as...**
  This allows you to save all the values you have entered into all the input screens of the program, so that you can start where you left off when running the program in future. You will be asked to supply a name for the file.

  – **Exit Program**
  This takes you out of the whole program.

● **Parts**

  – **About part 1/2**
  This gives a brief description of the part of the program you are currently in.

  – **Part 1: Finding Eigenvalues**
  This puts you into part 1 of the program.

  – **Part 2: Wavefunction Properties**
  This puts you into part 2 of the program.

Part 1: Finding Eigenvalues

This part of the program is used to find the eigenfunctions and eigenvalues of a number of different potential wells. Upon entering this part of the program the last three menu items become the following.

- ● Ang Mom

  - **Set Orbital Ang Mom, l**
    Asks you to enter an integer from 1 to 5 to specify the orbital angular momentum quantum number. The upper limit of 5 is considered enough for most purposes. (If you wanted the program to use higher values you could change it yourself, but you would have to write the code to calculate the Legendre polynomials for values of l greater than this.)

  - **Set Magnetic Ang Mom, m**
    Asks you to enter an integer from $-l$ to $+l$ to specify the magnetic quantum number.

  - **Display Spherical Harmonic**
    The program will draw a polar plot of the spherical harmonic with the values of l and m currently chosen (which you may observe from different polar angles), either as a three-dimensional representation or as a two-dimensional representation in the x-z ($\phi = 0$) plane.

- ● Potential

  - **Square Well**

  - **Parabolic**

  - **Coulombic**

  - **Lennard-Jones**

  - **Dumbell**

  - **Yukawa**

  - **User Defined**   (This item can only be accessed if the user has set the **userFlag** within the program itself, see section 6.4.2.)

  - **Vary Well Parameters**
    Allows you to change the parameters of the well. Choosing **[Ok]** or pressing <**Enter**> will accept whatever numbers appears in the boxes on the screen, unless they are outside the allowed ranges (which vary from well to well). If they are, the program will give an error message and wait for the numbers to be re-entered. Choosing **[Cancel]** or pressing <**Esc**> will abort entry and return to the main menu, with the well parameters unchanged from what they were before. Choosing **[View]** will display the lattice with the numbers just entered, but does not accept them permanently.

  - **Add a Perturbation**
    This allows the user to add a small perturbation to the well already selected. Choices available are—

    - * **Constant**

    - * **Linear**

    - * **Quadratic**

* **Cubic**

* **Quartic**

* **User Defined**   (This item can only be accessed if the user has set the **userFlag** within the program itself, see section 6.4.2.)

The coefficient which has to be entered must be small enough that the perturbation is nowhere more than 10% of the maximum depth of the well. Choosing **[Ok]** or pressing <**Enter**> will accept the choice of perturbation type and whatever number appears in the coefficient box on the screen, unless the number is too large. If it is, the program will give an error message and wait for the number to be re-entered. Choosing **[Cancel]** or pressing <**Esc**> will abort entry and return to the main menu, with the well unchanged from what it was before. Choosing **[View]** displays the lattice with the type and number just entered, but does not accept them permanently.

- **Method**
  This part of the program allows you to solve the Schrödinger equation for yourself. It is not strictly necessary if you are only interested in exploring physical results. It shows the methods by which those results were obtained, and will be useful if you make changes yourself to any of the program's parameters. It allows four different submenu choices:

  - **Try Energy (With Mouse)**
    This allows you to select an energy by clicking within the graph. The Schrödinger equation is then solved and the solution drawn on the screen. The value of the energy chosen also appears in a message box at the top of the screen, together with the number of nodes. *Note: This number might be one greater than the number of nodes you can actually see. That means there is another node off to the right somewhere.*

    Three hot keys may be selected, with the mouse or by pressing—

    * **Clear**   Redraw the screen without the previous solutions shown.

    * **Zoom In/Out**   Increase/decrease the scale by a factor of two, useful when trying to home in on an eigenvalue.

    * **Menu**   Return to the main menu.

  - **Try Energy (From Keyboard)**
    This allows you to enter an energy directly. The Schrödinger equation is then solved and the solution drawn on the screen. The value of the energy chosen also appears in a message box at the top of the screen, together with the number of nodes. Note: This number might be one greater than the number of nodes you can actually see. That means there is another node off to the right somewhere. This is important when you enter a number which is only approximately equal to an eigenvalue. The shape of the function may look as you think it should, but since the "real" eigenvalue may be a little less than your choice, the program will find one more node than you expect (which may well be off-screen).

Special results occur when the following choices are made.

* **[Ok]**   Accept the number entered and show the solution, as just described.

* **[View]**   Redraw the screen without the previous solutions shown.

* **[Cancel]**   Return to the main menu.

– **Hunt for Zero**
This allows you to find accurate eigenvalues by having the program hunt in a binary search between two limits. A complete description of how it works may be found in chapter 2 section 2.5.1.

– **Examine Solution**
This allows you to read from the screen actual values of the wave function and the x position, simply by pointing to the drawn curve with the mouse. The information appears in a message box at the top of the screen. Pressing or clicking on <Esc> returns you to the main menu. *Note: This menu item can only be selected immediately after having found a solution with* **Hunt for Zero** *or in a couple of other cases when there is unambiguously a wave function drawn on the screen capable of being examined.*

It allows you to read from the screen actual values of the wave function and the x position, simply by pointing to the drawn curve with the mouse. The information appears in a message box at the top of the screen. Pressing or clicking on <Esc> returns you to the main menu.

● **Spectrum**

– **Find Eigenvalues**
The program automatically searches for the eigenvalues. It does this by a recursive procedure, so they appear on the screen in no fixed order. As each is found, the energy level is drawn on the graph and a beep is sounded. This sound may be disabled if it annoys (see **Sound** below).

– **See Wave Functions**
This item displays the eigenfunctions previously found for the particular potential (and the value of l) that has been chosen. On first entry it asks you which of the eigenfunctions you want to see, thus:

Choose a level number ...   [Clear] [Ok] [Cancel]

Enter from the keyboard. If you try to give the program a number greater than the number of eigenvalues, or if you enter an alphabetic character, it will beep at you. When you have entered an acceptable number it will draw that eigenfunction, plot the probability, and display in the window the level number and the energy eigenvalue. It will also highlight on the potential plot the energy level in question. It will then wait for you to enter another level number. When you've seen enough eigenfunctions, choose **[Cancel]** or press <Esc> and you will return to the main menu.

– **Examine Solution**
This allows you to read from the screen actual values of the last eigenfunction you plotted and the corresponding x positions, simply by pointing to the

drawn curve with the mouse. The information appears in a message box at the top of the screen. Pressing or clicking on <**Esc**> returns you to the main menu. *Note: This menu item can only be selected immediately after having displayed one or more eigenfunctions with the previous two menu items or after* **Hunt for zero**, *when there also is unambiguously a wave function drawn on the screen capable of being examined.*

– **See Wfs and Probs**
This does the same as the previous menu item except that it will plot the probability density as well as the wave function each time.

– **Probability Cloud**
This item displays the probability of finding the particle at any point on the x-z ($\phi = 0$) plane for any of the eigenfunctions previously found for the particular potential (and the value of l) that has been chosen. On first entry it asks you which of the eigenfunctions you want to see, in the usual way.

Enter from the keyboard. If you try to give the program a number greater than the number of eigenvalues, or if you enter an alphabetic character, it will beep at you. When you have entered an acceptable number it will start putting randomly chosen points on the screen, chosen by using the probability density calculated. Thus a picture of the probability cloud will be built up.

The program will not stop of its own accord. You must choose one of four hot keys, with the mouse or by pressing—

* **Run/Stop**   Toggle between running and stopping.

* **Faster/Slower**   Increase/decrease the speed by a factor of two.

* **Restart**   Clear the screen and begin plotting again immediately.

* **Menu**   Return to the main menu.

– **Sound**
This toggles the sound on and off.

● **Units**

– **Units**
This allows you to choose from the following well shapes:

* **Atomic Units**   Choose to measure lengths in nm and energies in eV, masses in multiples of the electron mass.

* **Molecular Units**   Choose to measure lengths in nm and energies in eV, masses in multiples of the a.m.u.

* **Nuclear Units**   Choose to measure lengths in fm and energies in MeV, masses in multiples of the a.m.u.

– **Mass**
Choose the mass of the particle being investigated as a multiple of the mass unit currently chosen.

- **r Axis**
  Choose the maximum value or r along the axis, in the length units currently chosen.

Part 2: Wave Function Properties

This part of the program can be used to calculate various overlap integrals involving the eigenfunctions you found in part 1, or linear combinations of them. Note especially that you must be sure that you have calculated the eigenfunctions before attempting to use these features (unless you have not changed the potential from its default form). Upon entering this part of the program the last four menu items become the following.

- **Psi 1**
  This allows you to choose the wave function which will be the first item in the overlap integrand. There are two possible choices:

  - **Eigenstate, n = 1, 2, 3 . . .**
    Choose the eigenstate you want directly from the pull-down menu.

  - **General State**
    You are asked to supply coefficients for all of the bound state eigenfunctions for the well you are working with. None is allowed to be greater than 1.0000 If you enter any number greater than this, the program will give an error message and wait for you to re-enter a number in the right range. Choosing **[Cancel]** or pressing **<Esc>** will exit back to the main menu, leaving **psi1** unchanged.

- **Operator**
  This allows the operator in the middle of the overlap integrand to be specified. The choices are —

  $$1 \quad r \quad d/dr \quad r^2 \quad d^2/dr^2 \quad V \quad E \quad r.d/dr \quad d/dr.r \quad user1 \quad user2$$

  The operator currently in use is checked.

  The last two can only be accessed if the **userFlag** within the program itself has been set (see section 6.4.2).

- **Psi 2**
  This allows you to choose the wave function which will be the last item in the overlap integrand. There are three possible choices:

  - **Eigenstate, n = 1, 2, 3 . . .**
    Choose the eigenstate you want directly from the pull-down menu.

  - **General State**
    You are asked to supply coefficients for all of the boundstate eigenfunctions for the well you are working with. None is allowed to be greater than 1.0000. If you enter any number greater than this the program will give an error message and wait for you to re-enter a number in the right range. Choosing

[Cancel] or pressing <Esc> will exit back to the main menu, leaving **psi2** unchanged.

– **Some Other Function**
This allows you to enter a completely unrelated function. A screen will ask you to enter the algebraic form of the function. Only a limited number of forms are accepted. The default value appears is a "hat" function centered at $r = 0.075$ nm. Choosing [Ok] or pressing <Enter> accepts the algebraic form, provided it can be recognized by the CUPS parser. If it cannot, an error message will appear and the program will wait for further entry. Choosing [Cancel] or pressing <Esc> will exit back to the main menu, leaving **psi2** unchanged.

• **Integrate**
This will cause the overlap integration to be performed and the result shown in a separate box. If this menu item is selected before both **psi1** and **psi2** have been chosen, nothing will happen.

6.4.2   Possible Modifications to the Program

As you become more used to working with this program, you may find the need to do calculations that the author did not think of. The program is designed to make it easy for you to go inside and add extra pieces of code of your own.

You may of course change any part of the code you choose, but there are a number of "user defined" procedures already written which you may use as templates. These are gathered together in the part of the program headed by the comment

********* USER DEFINED PROCEDURES ********

In that part of the code is a Boolean variable **userFlag**. It is set to be *false*, but if you set it to it *true*, then at several points in the program an extra menu item will appear, called simply **User Defined**. By selecting that item you may access parts of the code which you yourself write.

There are three places where this occurs. In each case you are advised to write the specified procedure using the template that is there already, although you do not have to.

1. **Part 1.Potential**
If you write a procedure called **CalculateUserDVector**, you can incorporate your own well. Currently it calculates a simple exponential shape.

2. **Part 1.Potential|Add a Perturbation**
If you write a procedure called **SetUserPert**, you can add whatever perturbation you want to the potential. Currently it simply adds a constant times the orbital angular momentum to the potential.

3. **Part 2.Operator**
If you write two procedures called **ConstructUserKet1/2**, you can incorporate any two operations you want to be carried out on the function. Currently they simply reproduce the two operators r and d/dr.

Note that, if you *really* want to customize the program to your own needs, you may also change the name "user defined" which appears in the menu items to something of your own choosing. Just follow the templates in the procedure **SetUserNames**.

Here is a suggestion to show what you might do.

• Construct a nuclear potential well that is a simple exponential. This has an analytic solution for the Schrödinger equation, and it can be checked against the computed value. (See Merzbacher,[2] p. 211, problem 3.)

References

1. Schiff, L. I. *Quantum Mechanics*, 3rd ed. New York: McGraw-Hill, 1970.

2. Merzbacher, E. *Quantum Mechanics*, 2nd ed. New York: John Wiley & Sons, 1968.

3. Brehm, J. J., Mullin, W. J. *Introduction to the Structure of Matter*. New York: John Wiley & Sons, 1989.

4. Eisberg, R., Resnick, R. *Quantum Physics of Atoms, Molecules, Solids, Nuclei, and Particles*. New York: John Wiley & Sons, 1974.

7

Identical Particles in Quantum Mechanics

Daniel F. Styer

7.1 Introduction

In quantum mechanics, the wave function for a system of two particles in one dimension is a function of two variables: $\psi(x_1, x_2)$. (In this introduction we ignore spin.) If the two particles are identical, then the wave function must be either symmetric under the interchange of variables,

$$\psi(x_1, x_2) = +\psi(x_2, x_1), \tag{7.1}$$

or else antisymmetric under the interchange of variables,

$$\psi(x_1, x_2) = -\psi(x_2, x_1). \tag{7.2}$$

The foregoing assertion is an empirical rule that cannot be derived from any of the other principles of quantum mechanics. (Indeed there is currently considerable interest in *anyons*, hypothetical particles that obey all the principles of quantum mechanics except the interchange rule.[1])

The interchange rule has a number of surprising consequences. One is apparent through static wave functions: If the wave function is symmetric, the two particles tend to huddle together. If it is antisymmetric, they tend to spread apart. A separate but related consequence becomes apparent only as time evolves: If the wave function is symmetric, then the two particles—even though noninteracting—act much as if there is an attractive force between them. If it is antisymmetric, they act much as if there is a repulsive force between them. The major aim of program **Ident** is to illustrate and illuminate these consequences.

7.2 Spin, Wave Functions, and the Interchange Rule

This section considers the syntax of quantum mechanics: What is a legal expression for a wave function, and what does this expression mean? It begins with a very simple situation and progresses to more complex and subtle ones. You will find the beginning of this section to be very simple and familiar, but the later parts explain points that, in my experience, physics students and even professional physicists frequently misunderstand.

Notice, as you read this section, that it never mentions a Hamiltonian: all the statements are true for both interacting and noninteracting particles. This is because all the statements here consider only how to describe the system at a given instant in time, whereas the Hamiltonian affects only how systems evolve in time. (In other words, we are considering general states, not just energy eigenstates.)

7.2.1 One Particle

Consider a single particle of spin s moving in one dimension. Here s is a nonnegative integer or half-integer that depends only upon the type of particle under consideration: for pions s is always 0, for electrons it is always $\frac{1}{2}$, for deuterons it is always 1, etc. Then full information about the state of this system is contained in the configurational wave function $\Phi(x, m)$, where m, the projection of the spin on the z-axis, varies from $-s$ to $+s$ by integers. (This is not the only function that contains full information about the situation: see Exercise 7.5.) The interpretation of this wave function for, say, an electron, is that if the electron's spin projection and position are both measured, then the probability of finding it with spin up and located within an interval of width 0.001 centered on $x = 0.5$ is

$$|\Phi(0.5, +\tfrac{1}{2})|^2(0.001) , \tag{7.3}$$

while the probability of finding it with spin down in the same interval is

$$|\Phi(0.5, -\tfrac{1}{2})|^2(0.001) . \tag{7.4}$$

(In both cases—and in the parallel expressions that follow—we assume that the wave function varies negligibly over the width of the interval.)

In many important circumstances, *but not always*, the wave function can be factored into a spatial part and a spin part

$$\Phi(x, m) = \psi(x)\chi(m) . \tag{7.5}$$

[You may be more familiar with the analogous statement for wave functions of spin-0 particles in three dimensions: If $\psi(x, y, z)$ is the wave function, then in many but not all circumstances the wave function can be factored into $\psi(x, y, z) = \psi_A(x)\psi_B(y)\psi_C(z)$. For example, the energy eigenstates of the hydrogen atom cannot be factored in this way.]

7.2.2 Two Nonidentical Particles

Now consider the situation of two nonidentical particles, call them red and blue, moving in one dimension. Suppose the red particle has spin $s = 1$ and the blue one has spin $s = \frac{1}{2}$. Then the wave function

$$\Phi(x_1, m_1, x_2, m_2) \tag{7.6}$$

carries the interpretation that the probability of finding the red particle with spin projection $m = -1$ and located within an interval of width 0.001 centered on $x = 0.5$, and the blue particle with spin projection $m = +\frac{1}{2}$ and located within an interval of width 0.002 centered on $x = 0.3$, is

$$\left| \Phi(0.5, -1, 0.3, +\tfrac{1}{2}) \right|^2 (0.001)(0.002) \,. \tag{7.7}$$

(The particles are named red and blue rather than 1 and 2 to emphasize that the "red" labels a particle, whereas the "1," as in x_1 or m_1, labels a value of the position or of the spin projection.)

It is again possible that the wave function factorizes:

$$\Phi(x_1, m_1, x_2, m_2) = \psi(x_1, x_2)\chi(m_1, m_2) \tag{7.8}$$

or

$$\Phi(x_1, m_1, x_2, m_2) = \Phi_A(x_1, m_1)\Phi_B(x_2, m_2) \tag{7.9}$$

or even

$$\Phi(x_1, m_1, x_2, m_2) = \psi_A(x_1)\chi_A(m_1)\psi_B(x_2)\chi_B(m_2) \,. \tag{7.10}$$

But this is not required.

7.2.3 Two Identical Particles

Finally consider the situation of two identical particles moving in one dimension. The wave function has the same form as it did for two nonidentical particles, namely

$$\Phi(x_1, m_1, x_2, m_2) \,, \tag{7.11}$$

but it carries a very different interpretation. The probability of finding one particle with spin projection $m = -1$ and located within an interval of width 0.001 centered on $x = 0.5$, and the other particle with spin projection $m = 0$ and located within an interval of width 0.002 centered on $x = 0.3$, is

$$2\left| \Phi(0.5, -1, 0.3, 0) \right|^2 (0.001)(0.002) \,, \tag{7.12}$$

which is equal to

$$2\left| \Phi(0.3, 0, 0.5, -1) \right|^2 (0.002)(0.001) \,. \tag{7.13}$$

(Notice that it does not make sense to call one particle "#1" or "red" and the other "#2" or "blue"—the two particles are identical. But while the *particles* cannot be labeled, it is perfectly correct, and indeed it is necessary, to label the *positions*.) The prefactor 2 in these expressions arises to prevent double counting of probability (see Exercise 7.2).

This is the situation to which the interchange rule applies. The full wave function Φ is either symmetric or antisymmetric under the interchange of all the variables associated with a single particle, both space and spin:

$$\Phi(x_1, m_1, x_2, m_2) = \pm\Phi(x_2, m_2, x_1, m_1). \tag{7.14}$$

For this reason, the two probabilities (Eq. 7.12 and 7.13) are equal. Furthermore, the sign of the interchange symmetry, either $+$ or $-$, is governed only by the type of particle involved: for pions it is always $+$, for electrons it is always $-$. Particles for which the sign is always $+$ are called *bosons*, and those for which it is always $-$ are called *fermions*. It is an experimental fact that particles with integral spin s are bosons and those with half integral spin s are fermions.[2-4]

What does the interchange rule imply for those common situations in which the wave function factorizes into a spatial and a spin part? Suppose that

$$\Phi(x_1, m_1, x_2, m_2) = \psi(x_1, x_2)\chi(m_1, m_2). \tag{7.15}$$

If ψ and χ are both symmetric or both antisymmetric, then Φ is symmetric. If ψ is symmetric and χ is antisymmetric, or vice versa, then Φ is antisymmetric. Thus it is possible that the full wave function Φ is symmetric while the spatial wave function ψ is either symmetric or antisymmetric. The remainder of this chapter will treat only spatial wave functions, but it is important to remember that the full space-plus-spin wave function might not factorize, and that if it does, then the spatial wave function for two identical fermions may be either symmetric or antisymmetric under interchange of variables, depending on the symmetry of the spin wave function.

7.1 Exercise: Three Dimensions

Write down the condition that the interchange rule places on the wave function

$$\Phi(x_1, y_1, z_1, m_1, x_2, y_2, z_2, m_2) \tag{7.16}$$

of two identical fermions moving in three dimensions.

7.2 Exercise: Normalization Condition

The normalization condition for two identical spin-0 particles in one dimension is most insightfully written as

$$\int_{-\infty}^{+\infty} dx_1 \int_{x_1}^{+\infty} dx_2\, 2|\psi(x_1, x_2)|^2 = 1, \tag{7.17}$$

because this expression does not double count probabilities. (That is, it does not consider the situations "one particle at 0.5, one particle at 0.7" and "one particle at 0.7, one particle at 0.5" to be distinct.) Show that, for symmetric or antisymmetric wave functions, this is equivalent to the more usually encountered expression

$$\int_{-\infty}^{+\infty} dx_1 \int_{-\infty}^{+\infty} dx_2 |\psi(x_1, x_2)|^2 = 1. \tag{7.18}$$

7.3 Exercise: Spin-1 Particles

Two identical spin-1 particles move in one dimension.

a. Write down the normalization expression for the full wave function Φ.

b. What is the probability of finding one particle within an interval of width dx_1 centered on x_1, and the other within an interval of width dx_2 centered on x_2, without regard for spin projections? (This result is relevant to an experiment in which the positions of the particles are measured but their spin projections are not.)

7.4 Exercise: Nonsymmetric Spatial Wave Function?
One might argue that if

$$\psi(x_1, x_2) = i\psi(x_2, x_1) \tag{7.19}$$

and

$$\chi(m_1, m_2) = -i\chi(m_2, m_1), \tag{7.20}$$

then neither ψ nor χ would be symmetric or antisymmetric, yet their product Φ would be symmetric. Show, however, that the only function satisfying Eq. 7.19 for all values of x_1 and x_2 is $\psi(x_1, x_2) = 0$.

7.5 Exercise: The Interchange Rule in Another Representation
Full information about the state of two identical particles is contained not only in the configurational wave function $\Phi(x_1, m_1, x_2, m_2)$ but also in the momentum space wave function

$$\tilde{\Phi}(p_1, m_1, p_2, m_2) = \frac{1}{2\pi\hbar} \int_{-\infty}^{+\infty} dx_1 \int_{-\infty}^{+\infty} dx_2 \Phi(x_1, m_1, x_2, m_2) e^{-i(p_1 x_1 + p_2 x_2)/\hbar}.$$

$$\tag{7.21}$$

Show that if one representation is symmetric (or antisymmetric), then the other one is as well.

7.3 Building Symmetrized and Antisymmetrized Functions

If the spatial and spin parts of the wave function factorize, then each part is either symmetric or antisymmetric under the interchange of its two variables. It is therefore worthwhile to have a process for producing such functions.

Given an arbitrary two-variable function $f(x_1, x_2)$, you may construct from it a symmetric function

$$f(x_1, x_2) + f(x_2, x_1), \tag{7.22}$$

and an antisymmetric function

$$f(x_1, x_2) - f(x_2, x_1). \tag{7.23}$$

In this context the original function is called the "nonsymmetrized" function while the constructed functions (7.22 and 7.23) are called the "symmetrized" and "antisymmetrized" functions respectively. If the nonsymmetrized template function is a normalized quantum mechanical wave function $\psi(x_1, x_2)$, then the *normalized* constructed wave functions are (see Exercise 7.8)

$$\frac{\psi(x_1, x_2) \pm \psi(x_2, x_1)}{\sqrt{2 \pm Z_{sa}}}, \tag{7.24}$$

where

$$Z_{sa} \equiv 2 \int_{-\infty}^{+\infty} dx_1 \int_{-\infty}^{+\infty} dx_2 \psi^*(x_1, x_2)\psi(x_2, x_1). \tag{7.25}$$

7.6 Exercise: Decomposing Functions
Show that any function of two variables can be written as the sum of a symmetric function and an antisymmetric function.

7.7 Exercise: Symmetrizing the Symmetric
Suppose that the nonsymmetrized template function happens to already be symmetric or antisymmetric. (This explains why that function is called "nonsymmetrized" rather than "nonsymmetric.") What happens when you attempt to build symmetric and antisymmetric functions from it? What happens if the nonsymmetrized function is a wave function and expression 7.24 appears guaranteed to produce two normalized wave functions as output?

7.8 Exercise: Normalization

a. Verify expression 7.24.

b. In some circumstances, such as in finding energy eigenstates for non-interacting identical particles, the nonsymmetrized wave function is a product

$$\psi(x_1, x_2) = u_A(x_1)u_B(x_2), \tag{7.26}$$

where $u_A(x)$ and $u_B(x)$ are orthogonal. Show that in such circumstances the constant Z_{sa} vanishes.

7.9 Exercise: Going Backward
Given an arbitrary symmetric function $s(x_1, x_2)$ find at least one template function $f(x_1, x_2)$ such that $f(x_1, x_2)$ will build $s(x_1, x_2)$ through the symmetrization process of Eq. 7.22. Similarly for any given antisymmetric function.

7.10 Exercise: Corresponding Functions
The two functions 7.22 and 7.23 are in some sense corresponding, because both were constructed from the same building material. An obvious question is whether there is a unique antisymmetric function that "corresponds" to a given symmetric function. Such an antisymmetric function would be constructed by going backwards (as in the previous exercise) from the given symmetric function to a nonsymmetrized template $f(x_1, x_2)$, and then from that template to an antisymmetric function through Expression 7.23.

Show that this process does not produce a unique antisymmetric function and that, in fact, it can produce any antisymmetric function at all!

7.11 Exercise: Spin Wave Functions

This section has emphasized the symmetrization or antisymmetrization of spatial wave functions, in which the arguments are continuous variables running from $-\infty$ to $+\infty$. Produce results analogous to those in expressions 7.24 and 7.25 for the symmetrization and antisymmetrization of spin wave functions, in which the arguments are discrete variables running from $-s$ to $+s$ by integers.

7.4 *Reduced Probability Densities*

If $\psi(x_1, x_2)$ is the wave function for two particles (perhaps identical, perhaps not) in one dimension, then the probability density

$$\text{PD}_{\text{joint}}(x_1, x_2) = |\psi(x_1, x_2)|^2 \tag{7.27}$$

contains complete information about where the two particles are likely to be found if their positions are measured. (Note that it does *not* contain complete information about the state: it says nothing, for example, about the mean momenta, and it cannot be used to predict the future state of the system.) You might think that complete information is just the right amount, but in fact it is usually difficult to interpret this joint probability density—there is just too much information there to be readily assimilated. For that reason we introduce here two *reduced* probability densities: one-variable densities that contain less than complete information but that are easier to understand and interpret. (The reduced probability densities are useful for another purpose as well: It is psychologically comforting to see a one-dimensional situation described by a one-variable function.)

The first is the so-called single probability density. Whereas the joint probability density tells you the probability of finding one particle near one position and the other near some other position, the single probability density, $\text{PD}_{\text{single}}(x)$, tells you only the probability of finding a particle near one position without regard for where the other one is. To illustrate: if you performed many experiments on the same state, in each case finding the positions of both particles and recording both positions on a card, then from your data you would be able to find the joint probability density. If you cut each card in half and then jumbled them so that you could no longer tell which two positions were found in the same experiment, then you would have exactly enough information left to find the single probability density. In equations,

$$\text{PD}_{\text{single}}(x) = \frac{1}{2}\left[\int_{-\infty}^{+\infty} dx_1\, \text{PD}_{\text{joint}}(x_1, x) + \int_{-\infty}^{+\infty} dx_2\, \text{PD}_{\text{joint}}(x, x_2)\right]. \tag{7.28}$$

The factor of $\frac{1}{2}$ is necessary because $\text{PD}_{\text{joint}}(x_1, x_2)$ involves two particles, while $\text{PD}_{\text{single}}(x)$ involves only one. (In terms of the illustration, it is because the number of cards doubles when each one is cut in half.)

The single probability density is a very natural reduction of the joint probability density, but it leaves out all information about *correlations*. Suppose the first particle was positioned with a uniform density, but that the second one was always exactly 0.3 nm to its right. This very striking fact would be lost in the single probability density, which would be uniform. However, it would show up immediately through the so-called separation probability density, $PD_{sep}(s)$, which is defined so that the probability of finding the two particles separated by a distance between s and $s + ds$ is $PD_{sep}(s)ds$. (Note that s is always positive or zero.) In terms of the joint probability density,

$$PD_{sep}(s) = \int_{-\infty}^{+\infty} dx_1\, PD_{joint}(x_1, x_1 + s) + \int_{-\infty}^{+\infty} dx_1\, PD_{joint}(x_1, x_1 - s). \quad (7.29)$$

In terms of the illustration above, it is just the probability density that would be obtained if, in each experiment, you recorded not the positions of the two particles but the distance between them.

The information conveyed through the separation probability density is usually more interesting than that conveyed through the single probability density. Program **Ident** is capable of showing either reduced probability density, but separation is the default.

[Note: Even for those with considerable experience in interpreting abstract mathematical expressions, there is an almost irrepressible urge to interpret the joint probability density $PD_{joint}(x_1, x_2)$—representing two particles in one dimension— as a function $PD(x, y)$—representing one particle in two dimensions. The best antidote for this temptation is to work the first five exercises below.]

7.12 Exercise: Normalizations
Verify that $PD_{single}(x)$ and $PD_{sep}(s)$ as defined above are normalized.

7.13 Exercise: Concrete Illustrations of the Probability Densities

a. Suppose that one electron is located (with certainty) at $x = +0.5$ nm and another is located (again with certainty) at $x = -0.2$ nm. Sketch the joint, single, and separation probability densities.

b. Repeat part a but suppose that the second particle is a neutron.

7.14 Exercise: Uniform Distribution
Two particles are equally likely to be found anywhere along a line segment of length L. Show that the mean separation between the two particles is $L/3$.

7.15 Exercise: Applying the Joint Probability Density
Suppose that two particles are constrained to fall between $-L$ and $+L$. Given a joint probability distribution $PD_{joint}(x_1, x_2)$, find integral expressions for each of the following quantities:

a. The probability of finding one particle between $-L$ and $-L/2$ and the other between 0 and L.

b. The probability of finding both particles between $L/2$ and L.

c. The probability of finding the two particles separated by a distance of $L/2$ or less.

d. The expected number of particles located between $-L/2$ and $+L/2$.

7.16   Exercise: Why Bother?

Instead of describing the system through the two-variable probability density $PD_{joint}(x_1, x_2)$, it would be much clearer to use two one-variable probability densities, one for each particle. In three sentences or fewer, explain why this cannot be done.

7.17   Exercise: Reduction From a Symmetric Joint Density

If the wave function in Eq. 8.27 is symmetric or antisymmetric under interchange, then the joint probability density $PD_{joint}(x_1, x_2)$ is symmetric. What simplifications does this introduce into the definitions of $PD_{single}(x)$ and $PD_{sep}(s)$?

7.5   Tour of Program Ident

Program **Ident** simulates the quantal time development of two noninteracting identical particles moving in a one-dimensional infinite square well. You may choose any of several initial nonsymmetrized wave functions from a menu, and view probability densities corresponding to that wave function or to the symmetrized or antisymmetrized wave functions built from it. The program is particularly useful for developing your intuition concerning the character of symmetric and antisymmetric wave functions.

This section describes the highlights of **Ident**; more detailed information is given in section 7.6, "Running Program **Ident** (Reference)."

7.5.1   Static Wave Functions

Ident can exhibit many interesting phenomena even without using its time development capabilities. When the program starts up, it shows the joint and separation probability densities corresponding to a (nonsymmetrized) bivariate Gaussian wave function,

$$\psi(x_1, x_2) = A_n \exp(-[X_1^2 - 2gX_1X_2 + X_2^2])e^{i(k_1x_1 + k_2x_2)}, \qquad (7.30)$$

where A_n is a normalization constant,

$$X_1 \equiv \frac{x_1 - \langle x_1 \rangle}{d_1}, \quad \text{and} \quad X_2 \equiv \frac{x_2 - \langle x_2 \rangle}{d_2}. \qquad (7.31)$$

(The significance of the parameters d_1 and d_2 is explored in Exercise 7.29.) This wave function may seem intimidating at first, but it is really quite simple. It comes about in this way: Begin with a rotationally symmetric Gaussian joint probability density centered on the origin,

$$PD_{joint}(x_1, x_2) = Ae^{-(x_1^2 + x_2^2)/\text{constant}}. \qquad (7.32)$$

Stretch it along an axis (not necessarily one of the coordinate axes; see Exercise 7.30) so that its contours become ellipses rather than circles. Finally translate it so that its center lies at $(\langle x_1 \rangle, \langle x_2 \rangle)$. The result is just the joint probability density associated with the wave function (Eq. 7.30). The wave function's phase is always chosen, in this program, to have $k_2 = -k_1$ so that the reference frame used by the program is the zero momentum (or center of mass) frame. There is nothing terribly significant about the exact form of this wave function: it was selected simply as an example of a wave function that does not factorize into one part dependent only upon x_1 times another part dependent only upon x_2.

Choose **Compare Symm. and Antisymm.** from the **Symmetry** menu to see the probability densities associated with the symmetrized and antisymmetrized wave functions arising from this bivariate Gaussian. You will notice that these two look precisely the same! This is because the two particles are far apart (in a sense made precise in Exercise 7.23). To examine a situation in which the symmetric and antisymmetric probability densities differ, choose **Bivariate Gaussian** from the **Wave func** menu and set the mean values of x_1 and x_2 to -0.05 nm and $+0.05$ nm respectively. This produces a wave function in which the two particles are quite close relative to their associated position uncertainties. Now you will see that the probability densities associated with the symmetric and antisymmetric wave functions are very different indeed. (You may also wish to examine the nonsymmetrized wave function by choosing **Symmetry | Nonsymmetrized**.) This illustrates the general principle mentioned in the introduction: if the (spatial) wave function for two identical particles is symmetric, the two particles tend to huddle together; if it is antisymmetric, they tend to spread apart. Notice that this information is readily available through both the joint and the separation probability densities, whereas the single probability densities (accessible through **Plot What**) tend to hide it.

It is useful, at this point, to examine some of the wave functions available through **Wave func | Energy Eigenstate** or **Wave func | Combination of Energy States** (see Fig. 7.1). Notice that the huddle together versus spread apart principle applies in all cases.

7.18 Exercise: Terminology for Avoidance Without Repulsion

The English language, which is rich in terms for most things, is poor in terms that describe the tendency of identical particles to bunch together even in the absence of an attractive interaction, or to avoid one another even in the absence of a repulsive interaction. (This is not surprising, considering that English was developed before quantum mechanics.) This chapter uses the phrases "huddle together" and "spread apart": alternatives include "crowd together" versus "scatter about," "flock together" versus "avoid each other," and "congregate" versus "disperse." Try to come up with another pair of phrases that accurately describes the correlation of noninteracting identical particles. (The author would be interested in hearing your ideas.)

7.5.2 Time-Varying Wave Functions

Any wave function in **Ident** can be set into motion by choosing **Run** from the menu. This isn't very exciting if the initial wave function is an energy

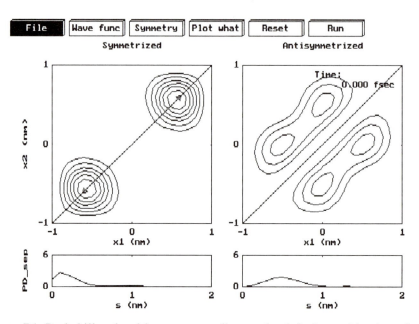

Figure 7.1: Probability densities corresponding to the default combination of energy eigenfunctions.

eigenfunction, because then the corresponding probability densities don't change at all. Reset the program to its initial condition by choosing **Reset | Reset to Defaults**, then start the wave function moving with **Run**. You are now watching two nonidentical particles collide—but the particles don't interact, so they pass right through each other! (This is not unique to quantum mechanics. Classical noninteracting particles behave in the same way. This is, in fact, the meaning of "noninteracting.") You can see this most easily by watching the separation probability density: Initially, it is very likely that the particles are far apart. Then the mound in $PD_{sep}(s)$ moves left toward the origin, crashes into zero (as the particles pass through each other) and finally moves away to the right again as the particles pull away from the collision. (If you keep on watching, you will see the particles bounce off of the edge walls and continue to move, but now in very complicated ways.)

Go back to the starting point with **Reset | Reset this Run**, but now set **Symmetry | Antisymmetrized**. Upon choosing **Run**, you will see the mound of $PD_{sep}(s)$ moving left, but it stops and begins to move right again *before* it reaches the origin. The two particles never sit right on top of each other. I cannot emphasize enough that these are *noninteracting* particles, and that this apparent repulsion is just that—apparent. It is not due to any force, it is not due to any term in the Hamiltonian; it is due only to the antisymmetrization requirement of the interchange rule.

Run the same collision over again with **Symmetry | Symmetrized**. You may be able to see that now there is an effective attraction between the two noninteracting identical particles, although this is more difficult to detect. The best way to see it is by starting over again with **Symmetry | Compare all Three** and with **Plot What | Show Mean Separation**. (See also Exercise 7.28.)

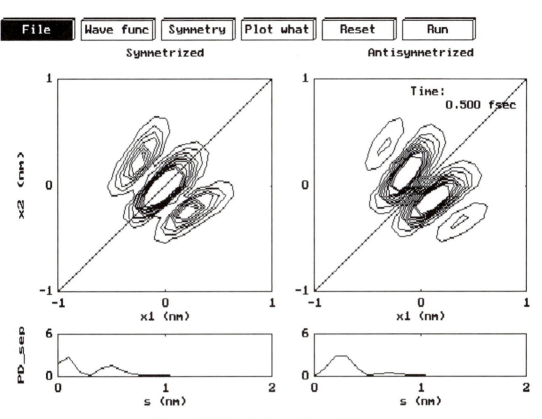

Figure 7.2: Identical particles colliding.

7.6 Running Program Ident (Reference)

7.6.1 Menu Items

- **File**: Documentation, control, and bookkeeping items.

  - **About CUPS**: Information about the CUPS project.

  - **About Program**: Information about program **Ident**.

  - **Things to Notice**: Points out several physics matters. Intended mainly for users who never bother to read documentation, and you are obviously not one of those!

  - **Configuration**: Modify colors, temporary directories, and other program characteristics.

  - **Open ...**: Read a profile file and set the current profile accordingly. The "profile" information consists of the initial nonsymmetrized wave function plus the symmetry, display, and animation speed settings.

  - **Save as ...**: Save the current profile information in a file.

– **Log Mean Separation**: Log mean values of the separation, $\langle s(t)\rangle$, to a file. Saving commences when **Run** is next chosen, and it stops when that run ends.

– **Exit Program**: Leave the program.

● **Wave func**: Set the initial nonsymmetrized wave function.

– **Bivariate Gaussian**: Set the initial wave function to a bivariate Gaussian (Eq. 7.30). An input screen pops up asking you to enter the parameters for the wave function.

– **Energy Eigenstate**: Set the initial wave function to an energy eigenstate of the "two noninteracting particles in a box" problem, namely

$$u_{n,m}(x_1, x_2) = \left[\frac{1}{\sqrt{L}}\sin\left(n\pi\frac{x_1 + L}{2L}\right)\right]\left[\frac{1}{\sqrt{L}}\sin\left(m\pi\frac{x_2 + L}{2L}\right)\right], \quad (7.33)$$

where the infinite square well is of width $2L$ centered on $x = 0$. An input screen pops up asking which eigenstate you want.

– **Combination of Energy States**: Set the initial wave function to a linear combination of three energy eigenfunctions, namely

$$a_A u_{2,1}(x_1, x_2) + a_B u_{1,3}(x_1, x_2) + a_C u_{3,2}(x_1, x_2). \quad (7.34)$$

An input screen pops up asking for the values of a_A, a_B, and a_C. (The particular values of n and m used here are in no way special. They can easily be changed by altering procedure **BuildWFCombo**.)

– **User Defined**: If you want to study an initial wave function that is not built in, you may code it yourself into **Ident**'s procedure **UserDefinedWF**. In the original program, this procedure generates a simple, pure real bivariate Gaussian wave function, but you may modify the procedure and recompile the program to get whatever you want.

● **Symmetry**: Choose the interchange symmetry of the wave function whose probability density is shown.

– **Nonsymmetrized**: Show the joint and a reduced probability density for the nonsymmetrized wave function.

– **Symmetrized**: Show the joint and a reduced probability density for the symmetrized wave function.

– **Antisymmetrized**: Show the joint and a reduced probability density for the antisymmetrized wave function.

– **Compare Symm. and Antisymm.**: Show the two above displays side by side.

– **Compare all Three**: Show a reduced probability density for all three wave functions: nonsymmetrized, symmetrized, and antisymmetrized. It would be nice to show the joint probability densities as well, but there simply isn't room on the screen!

- **Plot What**: Controls aspects not of what is computed, but of what is shown on the screen.

  - **Separation Probability Density**: The reduced probability density to be shown is the separation probability density (the default).

  - **Single Probability Density**: The reduced probability density to be shown is the single probability density.

  - **Show Mean Separation**: If separation probability density is shown, draw a vertical line marking the expectation value of the separation, $\langle s(t) \rangle$. If the line is present, it can be turned off by choosing **Show Mean Separation** a second time.

- **Reset**.

  - **Reset this Run**: Go back to the current initial wave function with the current symmetry and display settings.

  - **Reset to Defaults**: Go back to the situation at which the program starts up.

- **Run**: Start the time evolution.

7.6.2 Hot Keys

The hot keys for program **Ident** behave exactly as do those for program **QMTime**: see section 4.5.2.

7.6.3 Tips

- The two particles in program **Ident** have the mass of an electron. It would, however, be quite incorrect to say that they *are* electrons, because the particles in the simulation do not interact.

- When **Ident** starts up, it looks in the current working directory for a profile file named **IdtInit.pfl**. If it exists, the program reads that file and sets the profile accordingly. This is a good way to set up the program to suit your individual tastes. For more information on profiles and profile files, see the description of menu items **File I Open ...** and **File I Save as ...**.

- The presence of a working mouse can slow **Ident** by 50%. You may wish to turn your mouse off: simply refrain from loading the mouse driver.

7.7 Exercises

7.7.1 Running Ident

Ident invites exploration. Here are some suggestions to get you started.

7.19   Properties of Antisymmetric Wave Functions

Try out a variety of wave functions with **Symmetry | Antisymmetric** chosen. Notice that in all cases $PD_{joint}(x_1, x_2)$ vanishes if $x_1 = x_2$. Prove this fact analytically. How is it reflected in the separation probability density?

7.20   Probability of Zero Separation

Choose **Symmetry | Compare all Three** and observe the separation probability density at $s = 0$ for several different wave functions. From your observations, guess a relationship between $PD_{sep}(0)$ for symmetrized and nonsymmetrized wave functions and then verify (or modify) your guess analytically.

7.21   Visual Comparison

In order to develop familiarity with the features of symmetric and anti-symmetric wave functions, it is useful to observe and compare a large number of examples. In this context the time development capabilities of **Ident** can be thought of as a simple means to generate a large number of varied wave functions to be symmetrized or antisymmetrized. Accept the default initial wave function and choose **Symmetry | Compare Symm. and Antisymm.**. Choose **Run** and you will see many examples. What particular feature is present in the probability density for the symmetric wave function at time 2.8 fs?

7.22   Inequality for Mean Separations

Choose **Symmetry | Compare all Three** and **Plot What | Show Mean Separation** and observe a number of different wave functions: some energy eigenstates and some produced by the time evolution of a bivariate Gaussian.

a. Conjecture an inequality relating the mean separations $\langle s \rangle_N$, $\langle s \rangle_S$, and $\langle s \rangle_A$, for the nonsymmetrized, symmetrized, and antisymmetrized wave functions respectively.

b. Select an initial bivariate Gaussian wave function in which all the parameters vanish except for $\Delta x_1 = 0.10$ nm and $\Delta x_2 = 0.11$ nm. Time evolve this wave function for about 2 fs. Comment upon your conjecture.

c. (This part involves modifying the program code.) Program **Ident** comes with a sample user defined wave function procedure called **User-DefinedWFSample**. Implement this sample as your user defined wave function, then recompile and run the program. Comment again upon your conjecture.

7.23   When is (Anti)symmetrization Significant?

Choose **Wave func | Bivariate Gaussian** always using the default values for the positions, uncertainties, and momenta but varying the correlation coefficient g from $+0.9$ to -0.9 by steps of 0.2. At each step observe the situation first with **Symmetry | Nonsymmetrized** and then with

Symmetry I Compare Symm. and Antisymm. Notice that if the non-symmetrized $PD_{joint}(x_1, x_2)$ is negligible near the line $x_1 = x_2$, then the joint probability densities are very similar for both the symmetrized and antisymmetrized wave functions.

a. In this situation, compare the reduced (both separation and single) probability densities for the nonsymmetrized, symmetrized, and antisymmetrized wave functions.

b. In this situation, how is the joint probability density for the symmetrized wave function related to that of the nonsymmetrized function?

c. Prove analytically that this situation always arises if, in the nonsymmetrized case,

$$PD_{joint}(x_1, x_2) = P_A(x_1)P_B(x_2), \qquad (7.35)$$

where $P_A(x)$ and $P_B(x)$ are two functions whose product vanishes for all values of x.

7.24 Effects of (Anti)symmetrization on Energy Eigenstates

Turn on the display of mean separations using **Plot What I Show Mean Separation**. Choose **Symmetry I Compare all Three** and choose as wave function the sequence of energy eigenstates with $n = 1$ and with $m = 2, \ldots, 5$. As m increases, do the correlations due to symmetrization and antisymmetrization become more or less important?

7.25 Time Invariant Separation Probability Densities

From the menu, set **Symmetry I Compare Symm. and Antisymm.** and then set **Wave func I Combination of Energy States** with $a_A = 1$, $a_B = -0.5\ i$, and $a_C = 0$. Choose **Run**, and notice that the *joint* probability densities change with time but that the *separation* probability densities do not! Explore other combinations and make a conjecture as to the circumstances under which this curious behavior takes place. Can you prove your conjecture correct analytically?

7.26 Two-Particle Ehrenfest Theorem?

The Ehrenfest theorem is a result from one-particle quantum mechanics which states that, in many circumstances, the mean position of a quantal particle moves almost the same as a classical particle would if it were placed in the same potential. For the free particle there is no "almost" in the theorem: The *mean position* of a quantal free particle changes in precisely the same way that the *position* of a classical free particle would.

a. Use **Ident** to demonstrate that this result does not generalize to the separation in a free two-particle system.

b. The particles in **Ident** are not absolutely free, because of the infinite walls. How can you tell that these walls are not significantly affecting your observations?

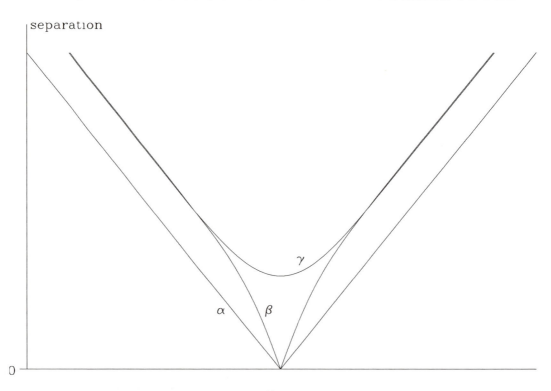

Figure 7.3: Classical interacting particles (Exercise 7.27).

7.27 Classical Interacting Particles

(This exercise is background for the next one.) In classical mechanics, of course, you needn't worry about nebulous entities like wave functions or probability densities: the particle separation is an ordinary real function of time.

a. Figure 7.3 presents three curves of separation versus time when two classical particles of the same mass approach each other in one dimension. In one case the two particles don't interact, in another the interaction is purely attractive, and in the third it is purely repulsive. Which is which? (The fact that the separation goes to zero in curves α and β may surprise you, but if two particles don't repel there is no reason they can't both occupy the same point at the same time.)

b. Sketch the separation versus time trajectory if the particle interaction is attractive at large distances and repulsive at small distances.

c. Show that the interaction force at a given time is proportional to the second derivative of the separation versus time trajectory, unless the separation vanishes. Is an attractive force positive or negative? (Hint: Consider the center of mass [zero momentum] reference frame.)

7.28 Effective Forces Between Noninteracting Particles

You know that the effective attraction (repulsion) between identical particles is not due to any force, but is a consequence of the (anti)symmetry requirement of the interchange rule. However, you might guess that it would be possible to form some sort of equivalent "effective force." Such an effective force would be a function of the mean separation $\langle s(t) \rangle$ and would be proportional to the "acceleration" $d^2 \langle s(t) \rangle / dt^2$. This idea can indeed lead to useful approximations, but this exercise shows (through a simple counterexample) that no such effective force can be exactly equivalent to the effects of the interchange requirement.

a. Use the default bivariate Gaussian wave function and **Symmetry | Symmetrized**. Choose **Plot What | Show Mean Separation** and **File | Log Mean Separation** to record $\langle s(t) \rangle$, and then **Run** for about 1 fs. Leave program **Ident** and use your favorite plotting program to plot an approximation for the acceleration as a function of mean separation.

b. Repeat part a for a bivariate Gaussian wave function with $\langle x_1 \rangle = -0.03$ nm. Are the two plots comparable?

c. Conclude that any "effective force" due to interchange symmetry must be a function of more than simply the mean separation.

7.7.2 Dissecting Ident

These exercises are designed to demonstrate and explain some of the many interesting calculations that went into writing program **Ident**. They do not require you to run the program: instead they encourage you to think about what went into writing the program.

7.29 Properties of the Bivariate Gaussian Density

Show that for the bivariate Gaussian density function

$$\mathrm{PD}(x_1, x_2) = A_n^2 e^{-2(X_1^2 - 2gX_1X_2 + X_2^2)}, \tag{7.36}$$

where

$$X_1 \equiv \frac{x_1 - \langle x_1 \rangle}{d_1} \quad \text{and} \quad X_2 \equiv \frac{x_2 - \langle x_2 \rangle}{d_2}, \tag{7.37}$$

a. normalization requires

$$A_n^2 = \frac{2\sqrt{1 - g^2}}{\pi d_1 d_2}; \tag{7.38}$$

b. the uncertainties are

$$\Delta x_1 = \frac{d_1}{2\sqrt{1 - g^2}} \quad \text{and} \quad \Delta x_2 = \frac{d_2}{2\sqrt{1 - g^2}}. \tag{7.39}$$

Hint: Convert to the variables (u_1, u_2), where

$$X_1 = u_1 - u_2 \quad \text{and} \quad X_2 = u_1 + u_2. \tag{7.40}$$

(In terms of these variables, the level curves of $PD(x_1, x_2)$ are ellipses with major and minor axes coincident with the u_1- and u_2-axes.)

7.30   The Bivariate Gaussian as a Stretched Mound

Suppose that the bivariate Gaussian wave function (Eq. 7.30) is centered on the origin, so that the joint probability density is

$$PD(x_1, x_2) = A_n^2 \exp\left(-2\left[\frac{x_1^2}{d_1^2} - 2g\frac{x_1 x_2}{d_1 d_2} + \frac{x_2^2}{d_2^2}\right]\right). \qquad (7.41)$$

The contours (or "level curves") of this density are a set of nested ellipses of the form

$$\frac{x_1^2}{d_1^2} - 2g\frac{x_1 x_2}{d_1 d_2} + \frac{x_2^2}{d_2^2} = \text{constant}. \qquad (7.42)$$

a. Write this equation in terms of the variables

$$x_1' = +\cos(\theta)x_1 + \sin(\theta)x_2 \qquad (7.43)$$

$$x_2' = -\sin(\theta)x_1 + \cos(\theta)x_2$$

where the (x_1', x_2') coordinate axes are simply rotated from the (x_1, x_2) axes by the angle θ.

b. Show that if

$$\tan(2\theta) = 4g\frac{d_1 d_2}{d_1^2 - d_2^2} \qquad (7.44)$$

then the (x_1', x_2') axes are coincident with the major and minor axes of the ellipses. Conclude that these are the axes mentioned in the text below Eq. 7.32 as "axes of stretching."

c. Show that there are four values of θ that satisfy Eq. 7.44, and interpret them in terms of axes of stretching or compressing.

7.31   Discretized Probability Densities

Suppose that $PD_{joint}(x_1, x_2)$ is known only at a finite set of grid points

$$PD_{joint}[j_1, j_2] \equiv PD_{joint}(x_{min} + (j_1 - 1)\Delta x, x_{min} + (j_2 - 1)\Delta x)$$

$$\text{for } j_1, j_2 = 1, 2, \ldots, N, \qquad (7.45)$$

and assume that $PD_{joint}(x_1, x_2)$ vanishes except when

$$x_{min} \leq x_1 \leq x_{max} \quad \text{and} \quad x_{min} \leq x_2 \leq x_{max}, \qquad (7.46)$$

where $x_{max} \equiv x_{min} + (N - 1)\Delta x$.

a. Show that a reasonable approximation for the single probability density is

$$PD_{single}(x_{min} + (j - 1)\Delta x) \approx PD_{single}[j]$$

$$\equiv \frac{\Delta x}{2}\left(\sum_{j_1=1}^{N} PD_{joint}[j_1, j] + \sum_{j_2=1}^{N} PD_{joint}[j, j_2]\right)$$

$$\text{for } j = 1, 2, \ldots, N. \qquad (7.47)$$

b. Show that a reasonable approximation for the separation probability density is

$$PD_{sep}((j-1)\Delta x) \approx PD_{sep}[j] \quad \text{for } j = 1, 2, \ldots, N, \quad (7.48)$$

where

$$PD_{sep}[1] \equiv \Delta x \sum_{j_1=1}^{N} PD_{joint}[j_1, j_1] \quad (7.49)$$

and

$$PD_{sep}[j] \equiv \Delta x \left(\sum_{j_1=1}^{N-j+1} PD_{joint}[j_1, j_1 + j - 1] \right.$$
$$\left. + \sum_{j_1=j}^{N} PD_{joint}[j_1, j_1 - j + 1] \right)$$
$$\text{for } j = 2, \ldots, N. \quad (7.50)$$

c. As j varies from 1 to N in Eq. 7.48, how does $s = (j-1)\Delta x$ vary, in terms of x_{min} and x_{max}?

d. Show that $PD_{sep}[1]$ is half of the value that would be calculated if you used expression 8.50 with $j = 1$.

e. Confirm that if the discretized joint probability density is normalized,

$$\sum_{j_1=1}^{N} \sum_{j_2=1}^{N} \Delta x^2 PD_{joint}[j_1, j_2] = 1, \quad (7.51)$$

then so are the discretized reduced probability densities.

7.8 Appendix: The Algorithm of Ident

Program **Ident** uses the Crank-Nicholson technique described in section 4.7 in conjunction with "operator splitting" (also called "alternating direction implicit" or ADI). This appendix provides an introduction to operator splitting; more details are available in section 9.3.2 and in Press et al.[5]

For one particle moving in one dimension, the Crank-Nicholson scheme is (see Eq. 4.44)

$$\left(1 + i\frac{\Delta t}{2\hbar} \mathbf{H} \right) \mathbf{\Psi}' = \left(1 - i\frac{\Delta t}{2\hbar} \mathbf{H} \right) \mathbf{\Psi}. \quad (7.52)$$

Here $\mathbf{\Psi}$ is the column matrix of discretized wave function values at time t and $\mathbf{\Psi}'$ is that matrix at time $t + \Delta t$. $\mathbf{H}$ is the square Hamiltonian matrix. Thus finding $\mathbf{\Psi}'$ involves inverting a matrix:

$$\mathbf{\Psi}' = \left(1 + i\frac{\Delta t}{2\hbar} \mathbf{H} \right)^{-1} \left(1 - i\frac{\Delta t}{2\hbar} \mathbf{H} \right) \mathbf{\Psi}. \quad (7.53)$$

In general matrix inversion is a computationally intensive process, but in this situation the task is much simplified because $\mathbf{H}$ is tridiagonal.

The paragraph above can be generalized directly to the case of two particles in one dimension (or the analogous case of one particle in two dimensions) except for the last sentence: $\mathbf{H}$ is no longer tridiagonal. One could resort to general matrix inversion techniques, but the program would then be intolerably slow. Instead, for free particles, we use operator splitting. In this technique the Hamiltonian is written as a sum of two parts,

$$\mathbf{H} = \mathbf{T}^{(1)} + \mathbf{T}^{(2)}, \tag{7.54}$$

where $\mathbf{T}^{(i)}$ is the kinetic energy matrix operating on the x_i variable (i.e., the discretization of the operator $-(\hbar^2/2m)\partial^2/\partial x_i^2$). Then the time-stepping (Eq. 7.52) is approximated by

$$\left(1 + i\frac{\Delta t}{2\hbar}\mathbf{T}^{(1)}\right)\left(1 + i\frac{\Delta t}{2\hbar}\mathbf{T}^{(2)}\right)\boldsymbol{\Psi}' = \left(1 - i\frac{\Delta t}{2\hbar}\mathbf{T}^{(1)}\right)\left(1 - i\frac{\Delta t}{2\hbar}\mathbf{T}^{(2)}\right)\boldsymbol{\Psi}, \tag{7.55}$$

which is solved in a two-step process through the introduction of an intermediate column matrix $\boldsymbol{\Psi}'$:

$$\left(1 + i\frac{\Delta t}{2\hbar}\mathbf{T}^{(1)}\right)\boldsymbol{\Psi}' = \left(1 - i\frac{\Delta t}{2\hbar}\mathbf{T}^{(2)}\right)\boldsymbol{\Psi} \tag{7.56}$$

$$\left(1 + i\frac{\Delta t}{2\hbar}\mathbf{T}^{(2)}\right)\boldsymbol{\Psi}' = \left(1 - i\frac{\Delta t}{2\hbar}\mathbf{T}^{(1)}\right)\boldsymbol{\Psi}'. \tag{7.57}$$

Some thought will demonstrate that both of these equations can be solved through the tridiagonal matrix trick. And that's all there is to operator splitting.

7.32   Exercise: Conservation Laws

Show that time development through operator splitting conserves both probability and energy. (Note that the matrices $\mathbf{T}^{(1)}$ and $\mathbf{T}^{(2)}$ commute.)

7.33   Exercise: Interchange Symmetry and Operator Splitting

a. Show that time development through operator splitting conserves interchange symmetry.

b. Furthermore, show that the same result is achieved whether a nonsymmetrized wave function is (anti)symmetrized and then stepped forward in time, or stepped forward and then (anti)symmetrized.

(These results are vital to program **Ident**, which time evolves only the nonsymmetrized wave function, and only symmetrizes or antisymmetrizes the evolved wave function immediately before displaying it.)

7.34   Exercise: Interacting Identical Particles?

It would be very desirable to change **Ident** to handle two *interacting* identical particles. However, it appears that when operator splitting is generalized for use in this situation, probability is not conserved. There is a published paper[6] that purports to produce a probability conserving generalization, but it contains an error. See if you can find it. (Hint: Look closely at the definition of $\hat{U}(\delta)$ two equations below equation (A4) on page 67 of the reference. What does A/B mean if A and B are matrices that do not commute?)

Acknowledgments

A preliminary version of program **Ident** was reviewed and critiqued by all members of the CUPS collaboration (as were all of the CUPS programs). Many of these reviews were helpful to me, but the comments and suggestions of Joseph Rothberg were particularly insightful and resulted in great improvements.

References

1. Canright, G. S., Girvin, S. M. Fractional statistics: quantum possibilities in two dimensions. Science **247**:1197–1205, 1990.

2. French, A. P., Taylor, E. F. *An Introduction to Quantum Physics*. New York: W.W. Norton and Company, 1978, chapter 13.

3. Lévy-Leblond, J.-M., Balibar, F. *Quantics: Rudiments of Quantum Physics*. Amsterdam: North-Holland, 1990, chapter 7.

4. Kekez, D., Ljubičić, A., Logan, B. A. An upper limit to violations of the Pauli exclusion principle. Nature **348**:224, 1990.

5. Press, W. H., Flannery, B. P., Teukolsky, S. A., Vetterling, W.T. *Numerical Recipes*. Cambridge: Cambridge University Press, 1986, sections 17.3 and 17.6.

6. Galbraith, I., Ching, Y. S., Abraham, E. Two-dimensional time-dependent quantum-mechanical scattering event. American Journal of Physics **52**:60–68, 1984.

8

Stationary Scattering States in Three Dimensions

John R. Hiller

8.1 Introduction

Scattering processes in three dimensions are of direct physical interest. They are the primary means by which experimentalists explore the nature of fundamental particles and their interactions. The experiments cover a wide range of scales, from the structure of crystals to the structure of protons. For example, the scattering of electrons from deuterons has been used to study (1) the attractive Coulomb force between the electron and the positively charged deuteron; (2) the effects of the parity-violating weak interaction[1]; (3) the binding of the proton and neutron that make up the deuteron, in both elastic and inelastic scattering, including tests of perturbative quantum chromodynamics[2]; and (4) the quark substructure of the proton and the neutron in very inelastic events.[3]

In the typical scattering experiment, a beam of particles is directed toward a target from which they may scatter to various angles. The probability for scattering to occur depends on the interaction between the beam particle and the target. The measurement of this probability is done by counting particles that enter detectors, and the results are reported in terms of the apparent area that the target presents to the beam.

This apparent area is called the *cross section*. To see its meaning in a classical context, consider a beach ball caught in a cloud burst. What number of rain drops actually strike the ball? Assume that the rain falls uniformly and vertically at a constant rate. The answer then depends on the flux of rain drops (the number of drops per unit horizontal area per unit time), the horizontal area covered by the ball, and the duration of the cloud burst. The product of these three things yields the number of drops scattered by the ball. The horizontal area covered by the ball is called its (total) cross section, and is equal to π times the square of its radius.

Just as in the case of the cloud burst, particle beams are much wider than the region in which scattering takes place, and the total cross section of the target determines the number of particles that will be scattered per unit time. However, in many experiments the detector will count only those particles scattered through a small range of angles. Only a fraction of the apparent area of the target is responsible for these scattering events. This fraction is used to define a *differential* cross section. When such a differential cross section is integrated over all possible scattering directions, the total cross section is recovered.

The challenge to theorists is to predict cross sections based on models for the particle interactions, or to use data already available to determine parameters in a model. The simplest approach is a nonrelativistic analysis based on the Schrödinger equation, in which a chosen potential models the target–beam interaction. Within this framework various directions can be taken.[4,5]

Here we will use partial wave analysis,[6-13] which is based on an expansion of the scattering wave function in terms of angular momentum wave functions. As discussed in chapter 6, each angular momentum wave function consists of a radial wave function and a spherical harmonic. Because we will analyze scattering from spherically symmetric potentials only, all the effects of the interaction are contained in the radial wave function. At large distances from the target, the only significant difference between the radial wave function of a scattered particle and one of an unscattered particle is a shift in phase. From these *phase shifts* the differential and total cross sections can be computed.

The program **Scattr3D** computes the individual radial wave functions and from them constructs the contributions to the complete scattering wave function. The effects of the phase shifts are summed over a finite range of angular momenta in order to obtain approximate differential and total cross sections.

The sections that follow provide some background in quantum scattering and describe partial wave analysis. The implementation in **Scattr3D** is also described.

8.2 *Quantum Scattering*

8.2.1 **Preliminaries**

We consider the elastic scattering of a uniform beam of monoenergetic particles from a target, the interaction with which is described by a potential $V(\mathbf{r})$. The beam axis is a convenient choice for the polar axis of a spherical coordinate system. The potential will be assumed to be spherically symmetric and to be negligible beyond a finite radius r_{max}.

The beam particle wave function is chosen to be of the form $\psi(\mathbf{r}) \exp(-iEt/\hbar)$, which represents a stationary state of definite energy.* Outside r_{max}, ψ is a combination of a plane wave that represents the incoming particle and a spherical wave that represents the scattered particle. The spherical wave is of the form $exp(ikr)/r$, which corresponds to an outgoing probability current. The

---

*See the discussions of stationary states in chapters 2 and 3. The extension to three dimensions is straightforward.

factor $1/r$ must be present to yield a finite probability of being in a spherical shell at infinite radius, and thereby reasonably represent particles scattered to infinity. Based on these ingredients, the amplitude ψ takes the form

$$\psi = e^{i\mathbf{k}\cdot\mathbf{r}} + f(\theta, \phi)\frac{e^{ikr}}{r}, \qquad (8.1)$$

where $k = \sqrt{2\mu E}/\hbar$, with μ the reduced mass, and $\mathbf{k} = k\hat{z}$. The function $f(\theta, \phi)$ is called the *scattering amplitude*.

Inside r_{max} the Schrödinger equation must be solved, subject to the requirement that the solution match the form of Eq. 8.1 at r_{max}. This is analogous to the process used in the one dimensional case, discussed in chapter 3, where the solutions at the exterior points in Eq. 3.16 are matched to a solution computed in the interior. The ratios B/A and C/A, which are then computable in the one-dimensional case, correspond here to $f(\pi, 0)$ and $f(0, 0)$, respectively. Because we now work in three dimensions, the amplitude of the scattered wave is not described by just two complex numbers but is instead a complex function of the direction from the scattering center.

Since an experiment involves counting particles that reach a detector, we need to be able to predict the fraction that will arrive at a detector placed at a particular angular position. As one might expect, by comparison with the one-dimensional problem, this number is proportional to $|f|^2$. It is also proportional to the size of the detector.

To measure the size of the detector, imagine a sphere, centered at the target, with a radius equal to the distance from the target. Every scattered particle will strike this sphere. The detector occupies some fraction dA of the area of this sphere, and it is that fraction that is the meaningful measure of its size. This fraction leads to the concept of solid angle; the detector is said to subtend a solid angle of $d\Omega \equiv 4\pi(dA/4\pi r^2) = dA/r^2$. A 4π detector is one that covers the entire sphere.

The number of particles recorded by a detector determines the differential cross section. If $\mathcal{N}(\theta, \phi)$ is the number of particles, per unit solid angle, that are scattered in the direction specified by θ and ϕ, then $\mathcal{N}\,d\Omega$ is the number that enter the detector. This number is some fraction of the total number of particles directed at the target. Let $\mathcal{N}_0$ be the total number of incoming particles per unit of cross-sectional area of the beam. The apparent area of the target for scattering into the detector is then given by the ratio of $\mathcal{N}\,d\Omega$ to $\mathcal{N}_0$. When divided by the solid angle of the detector, this ratio yields the differential cross section:

$$\frac{d\sigma}{d\Omega} = \frac{\mathcal{N}}{\mathcal{N}_0}. \qquad (8.2)$$

Once $d\sigma/d\Omega$ is known, the number of particles scattered to a detector of solid angle $\Delta\Omega'$ at (θ', ϕ') can be computed from $\mathcal{N}_0 \times \Delta\Omega' \times d\sigma(\theta', \phi')/d\Omega$. In practice, the number actually detected will depend on the efficiency of the detector. Experimental results reported in the literature usually include corrections for inefficiency and can be compared directly to the theoretical ideal.

The total cross section σ is just the total apparent area of the target. It can be determined from the number of particles scattered in any direction and is then simply the integral of the differential cross section:

$$\sigma = \int \frac{d\sigma}{d\Omega}\,d\Omega. \qquad (8.3)$$

Given σ, the total number of particles scattered is $\mathcal{N}_0 \times \sigma$.

The connection to $|f|^2$ must now be made. Outside r_{max} scattered particles travel with speed $\hbar k/\mu$ in the direction of the detector. The probability current density in that direction is, then, $(\hbar k/\mu)|(f/r)\exp(ikr)|^2$. The physical area of the detector is $r^2 d\Omega$. Therefore, the rate at which particles enter the detector is $(\hbar k/\mu)|f|^2 d\Omega$. The incoming current density of the beam is just $\hbar k/\mu$, which can also be considered the incoming flux. Each rate can be multiplied by the duration of the experiment to obtain particle counts, but the duration disappears in the ratio. The ratio of rates is

$$d\sigma = \frac{\frac{\hbar k}{\mu}|f|^2 d\Omega}{\frac{\hbar k}{\mu}} = |f|^2 d\Omega . \tag{8.4}$$

Thus we have

$$\frac{d\sigma}{d\Omega} = |f|^2 . \tag{8.5}$$

Clearly, the differential and total cross sections are directly measurable. They can also be predicted by theory. To predict them, one solves the Schrödinger equation at the specified energy, and, in the process of matching the boundary conditions (Eq. 8.1), extracts the scattering amplitude f.

8.2.2   Partial Wave Analysis

We seek a solution of the time-independent Schrödinger equation,

$$-\frac{\hbar^2}{2\mu}\nabla^2\psi + V(r)\psi = E\psi , \tag{8.6}$$

that satisfies the boundary conditions (Eq. 8.1) beyond the radius r_{max} at which V becomes negligible. The solution is to be built from the spherical harmonics[*] Y_{lm}, which are the eigenfunctions of the angular momentum operators L^2 and L_z,

$$L^2 Y_{lm} = l(l+1)\hbar^2 Y_{lm}, \quad L_z Y_{lm} = m\hbar Y_{lm} . \tag{8.7}$$

Because the beam is cylindrically symmetric, only functions with $m = 0$ will contribute; the others are dependent on the azimuthal angle ϕ. The spherical harmonics that remain are proportional to the Legendre polynomials $P_l(\cos\theta)$. Therefore, we write

$$\psi = \sum_{l=0}^{\infty} a_l \frac{u_{El}(r)}{r} P_l(\cos\theta) , \tag{8.8}$$

with a_l a constant coefficient and u_{El} a radial wave function. To complete the solution we must obtain the u_{El} and fit the boundary conditions.

---

[*]See the discussions of spherical harmonics and angular momentum, and of radial wave functions, in chapter 6.

Radial Wave Functions

On substitution of Eq. 8.8 into the Schrödinger equation (Eq. 8.6) and use of $\hbar^2 \nabla^2 = (\hbar^2/r)(\partial^2/\partial r^2)r - L^2$, we obtain

$$\sum_{l=0}^{\infty} a_l \left\{ \frac{-\hbar^2}{2\mu} \frac{d^2 u_{El}}{dr^2} + \left[\frac{l(l+1)}{2\mu r^2} + V(r) \right] u_{El} \right\} P_l(\cos\theta) = \sum_{l=0}^{\infty} E a_l u_{El} P_l(\cos\theta) \,. \quad (8.9)$$

The linear independence of the P_l leads immediately to individual radial equations that the u_{El} must satisfy:

$$-\frac{\hbar^2}{2\mu} \frac{d^2 u_{El}}{dr^2} + \left[\frac{l(l+1)}{2\mu r^2} + V(r) \right] u_{El} = E u_{El} \,. \quad (8.10)$$

At $r = 0$, we also require that u_{El} be zero, so that ψ is finite.

In a region where the potential is constant and less than the energy, the radial equation has the well known solutions $rj_l(k'r)$ and $rn_l(k'r)$, where j_l and n_l are the regular and irregular spherical Bessel functions[7,14] and $k' = \sqrt{2\mu(E - V)}/\hbar$. A general solution with angular momentum l is a linear combination of the two:

$$u_{El} = A_l r j_l(k'r) + B_l r n_l(k'r) \,. \quad (8.11)$$

If $r = 0$ is part of the region of interest, n_l is excluded due to its singular behavior, and its coefficient must be zero. In particular, if the potential is everywhere zero, then $k' = k \equiv \sqrt{2\mu E}/\hbar$ and $u_{El} = A_l j_l(kr)$.

This last comment implies that there are at least two ways to represent the infinitely degenerate states of a free particle in three dimensions. One is the simple plane wave $e^{i\mathbf{k}\cdot\mathbf{r}}$, where the direction of $\mathbf{k}$ can be varied, and the other is $j_l(kr)Y_{lm}(\theta,\phi)$, where l and m can be varied. The two sets of eigenfunctions must be related, and, indeed, this is the case. Since here we are concerned only with plane waves traveling in the z direction, we can consider the following special case:

$$e^{i(k\hat{z})\cdot\mathbf{r}} = \sum_{l=0}^{\infty} i^l(2l+1)j_l(kr)P_l(\cos\theta) \,, \quad (8.12)$$

This can be derived as a generalized Fourier series in the Legendre polynomials, on use of the integral representation for the Bessel functions,[15]

$$j_l(kr) = \frac{i^l}{2} \int_{-1}^{1} d(\cos\theta) e^{ikr\cos\theta} P_l(\cos\theta) \,, \quad (8.13)$$

and the orthonormality of the Legendre polynomials,

$$\int_{-1}^{1} d(\cos\theta) P_l(\cos\theta) P_{l'}(\cos\theta) = \frac{2}{2l+1} \delta_{ll'} \,. \quad (8.14)$$

The connection between plane waves and partial waves (Eq. 8.12) will prove useful below.

Phase Shifts

At large distances from the target there is a simple relationship between the radial wave function for the scattered partial wave and the radial wave function for a free particle of the same energy. To leading order in $1/r$ they differ only by a shift in phase. If the potential is repulsive, the scattered wave is pushed ahead of the free wave; if attractive, the scattered wave is pulled back.

This phase shift can be determined analytically by using the asymptotic forms of the spherical Bessel functions j_l and n_l. They are[15]

$$j_l(kr) \sim \frac{\sin(kr - l\pi/2)}{kr} + \mathcal{O}\left(\frac{1}{(kr)^2}\right) \tag{8.15}$$

and

$$n_l(kr) \sim -\frac{\cos(kr - l\pi/2)}{kr} + \mathcal{O}\left(\frac{1}{(kr)^2}\right). \tag{8.16}$$

The scattered partial wave in Eq. 8.11 is then asymptotically of the form

$$u_{El} \sim \frac{\sqrt{A_l^2 + B_l^2}}{k}\left[\frac{A_l}{\sqrt{A_l^2 + B_l^2}}\sin(kr - l\pi/2) - \frac{B_l}{\sqrt{A_l^2 + B_l^2}}\cos(kr - l\pi/2)\right]. \tag{8.17}$$

The factor $\sqrt{A_l^2 + B_l^2}$ has been introduced to permit the natural reparameterization of the coefficients as

$$A_l' = \sqrt{A_l^2 + B_l^2}, \quad A_l = A_l' \cos \delta_l, \quad B_l = -A_l' \sin \delta_l, \tag{8.18}$$

so that

$$u_{El} \sim \frac{A_l'}{k}\sin(kr - l\pi/2 + \delta_l). \tag{8.19}$$

One can now see directly from the comparison of the argument of this sine function with that of the sine in the free partial wave $rj_l(kr)$, that the only difference is the phase shift δ_l. It is to be computed from the coefficients A_l and B_l, which are obtained from the solution of the radial wave equation (Eq. 8.10), subject to the condition $u_{El}(0) = 0$. Notice that the phase shift is in general energy dependent.

To simplify later analysis, we set A_l' to one. This amounts to a redefinition of the coefficient a_l in the partial wave expansion (Eq. 8.8). With this adjustment, the form of u_{El} beyond $r_{\max}$ is required to be

$$u_{El} = r[\cos \delta_l j_l(kr) - \sin \delta_l n_l(kr)] \quad \text{for } r > r_{\max}. \tag{8.20}$$

This condition determines both the phase shift and the normalization of u_{El}.

Application of the Boundary Condition

The solution to the Schrödinger equation has been explored sufficiently to permit a comparison to the boundary condition (Eq. 8.1). From this comparison we will extract the coefficients a_l and the scattering amplitude f in terms of the phase shifts.

To facilitate comparison, we write the boundary condition in the form of a partial wave expansion. The plane-wave piece is expanded as in Eq. 8.12. For the spherical piece, we need only write an expansion of the scattering amplitude,[*]

$$f_k(\theta) = \sum_{l=0}^{\infty} (2l + 1)f_l(k)P_l(\cos \theta), \tag{8.21}$$

---

*Almost any function of θ can be expanded in terms of Legendre polynomials.

with coefficients f_l to be determined. The boundary condition then takes the form

$$\psi \sim \sum_{l=0}^{\infty} \left[i^l j_l(kr) + f_l(k)\frac{e^{ikr}}{r} \right] (2l + 1)P_l(\cos \theta) . \tag{8.22}$$

Use of the asymptotic form for j_l yields

$$\psi \sim \sum_{l=0}^{\infty} \left[i^l \frac{\sin(kr - l\pi/2)}{kr} + f_l(k)\frac{e^{ikr}}{r} \right] (2l + 1)P_l(\cos \theta) . \tag{8.23}$$

This must be matched by the asymptotic form of the sum over solutions to the radial wave equation, which from substitution of Eq. 8.19 into Eq. 8.8 must be*

$$\sum_{l=0}^{\infty} a_l \frac{\sin(kr - l\pi/2 + \delta_l)}{kr} P_l(\cos \theta) . \tag{8.24}$$

On equating coefficients of the P_l, we obtain

$$a_l \frac{\sin(kr - l\pi/2 + \delta_l)}{kr} = i^l(2l + 1)\frac{\sin(kr - l\pi/2)}{kr} + (2l + 1)f_l(k)\frac{e^{ikr}}{r} . \tag{8.25}$$

The sine functions can be written as linear combinations of e^{ikr} and e^{-ikr}. Because these exponential functions are independent, we can solve for both f_l and a_l to find, after some algebra,

$$f_l(k) = \frac{e^{i\delta_l} \sin \delta_l}{k}, \quad a_l = i^l(2l + 1)e^{i\delta_l} . \tag{8.26}$$

The phase shifts remain as the only undetermined quantity, and can only be obtained after the potential is specified and the wave functions u_{El} computed.

Partial Wave Expansions

The expansions for the wave function and the scattering amplitude can now be written in terms of the phase shift:

$$\psi = \sum_{l=0}^{\infty} (2l + 1)i^l e^{i\delta_l}\frac{u_{El}(r)}{r}P_l(\cos \theta) , \tag{8.27}$$

$$f_k(\theta) = \frac{1}{k} \sum_{l=0}^{\infty} (2l + 1)i^l e^{i\delta_l} \sin \delta_l P_l(\cos \theta) . \tag{8.28}$$

The differential and total cross sections are then also immediately determined

$$\frac{d\sigma}{d\Omega} = |f_k(\theta)|^2 = \frac{1}{k^2} \sum_{l,l'=0}^{\infty} (2l + 1)(2l' + 1)e^{i(\delta_l - \delta_{l'})} \sin \delta_l \sin \delta_{l'} P_l(\cos \theta)P_{l'}(\cos \theta) , \tag{8.29}$$

$$\sigma = \int \frac{d\sigma}{d\Omega}d\Omega = \frac{4\pi}{k^2} \sum_{l=0}^{\infty} (2l + 1)\sin^2\delta_l . \tag{8.30}$$

---

*The coefficient A_l^l that appears in Eq. 8.19 has been set to one, as discussed previously.

In the case of the total cross section the orthogonality of the Legendre polynomials, expressed in Eq. 8.14, has been used. Notice that, up to this point, everything is exact.

Two approximations can now be made to make calculations practical. One is the use of approximate solutions for u_{El}, in particular, numerical solutions. The other is a truncation of the sums over l.

For large l, a classical approximation applies. The particle can be considered to travel a definite trajectory, which begins along a line parallel to the z axis. The initial distance b from the z-axis is called the impact parameter. The relationship between the impact parameter and the angular momentum $\hbar l$ is

$$\hbar k b = \hbar l, \tag{8.31}$$

where $\hbar k$ is the incoming momentum. If l is such that b is greater than r_{max}, this partial wave will not encounter the potential V to any significant extent, nor will any partial wave of higher l. Therefore, an upper bound l_{max} can be set by the definition

$$\hbar k r_{max} = \hbar l_{max}. \tag{8.32}$$

For low-energy scattering, k is small and only small l will contribute. In fact, one frequently finds that the S and P waves ($l = 0$ and 1), or even just the S wave, are enough. Notice that S-wave scattering is isotropic; because $P_0 = 1$, the amplitude is independent of the scattering angle.

Experimental results are frequently parameterized in terms of phase shifts. The δ_l are extracted from fits of $d\sigma/d\Omega$ to data, under the assumption that only a few l contribute. Such an assumption can be checked experimentally by seeing how much the fit improves when new partial waves are added.

Optical Theorem

The partial wave expansion can be used to prove the optical theorem, which is a proportionality between the total cross section and the imaginary part of the forward scattering amplitude. Specifically, the relationship is

$$\sigma = \frac{4\pi}{k} \mathcal{I}m f_k(0). \tag{8.33}$$

This equality follows directly from Eqs. 8.30 and 8.28, and the fact that $P_l(1) = 1$.

8.2.3   Piecewise-Constant Potentials

As in the case of one-dimensional scattering, piecewise-constant potentials admit analytic solutions. However, the functions involved are Bessel functions rather than simple exponentials. Linearly independent solutions of the Schrödinger equation in a region where the potential is constant, and less than the energy, have already been given as r times the spherical Bessel functions $j_l(k'r)$ and $n_l(k'r)$ with $k' = \sqrt{2\mu(E - V)}/\hbar$. When the energy is less than the constant value of the potential, the solutions may be chosen to be r times the spherical modified Bessel functions $i_l(\kappa r)$ and $k_l(\kappa r)$, with $\kappa = \sqrt{2\mu(V - E)}/\hbar$. Not surprisingly, these functions are related to ordinary spherical Bessel functions of imaginary argument[15]:

$$i_l(x) = i^l j_l(ix), \quad k_l(x) = -i^l[j_l(ix) + i n_l(ix)]. \tag{8.34}$$

When the energy is equal to the potential, one can readily show that the independent solutions are r^{l+1} and r^{-l}. The coefficients of these functions are then determined by matching conditions at each point of discontinuity in the potential.

The calculation of the coefficients can follow a pattern analogous to that used in chapter 3 for the one-dimensional case. One key difference, however, is that the interval closest to the origin is where the shape of the wave function is known, and the calculation must proceed from there to the outermost interval $[r_{max}, \infty]$. To illustrate the process of finding the coefficients and the phase shifts, we consider some simple examples.

Hard Sphere

The simplest situation is one where the potential is infinite inside some radius a and zero elsewhere. There is then only one region in which to seek the wave function, and there it must have the form

$$u_{El} = A j_l(kr) + B n_l(kr). \tag{8.35}$$

The condition at the discontinuity in the potential is simply $u_{El}(a) = 0$, which implies that

$$\frac{B}{A} = -\frac{j_l(ka)}{n_l(ka)}. \tag{8.36}$$

Thus the coefficients are determined up to a normalizing factor. This factor, and the phase shift, are determined by the condition in Eq. 8.20, which becomes

$$A\left[j_l(kr) - \frac{j_l(ka)}{n_l(ka)} n_l(kr) \right] = \cos \delta_l j_l(kr) - \sin \delta_l n_l(kr). \tag{8.37}$$

Because j_l and n_l are linearly independent, we can conclude that

$$\tan \delta_l = \frac{j_l(ka)}{n_l(ka)} \quad \text{and} \quad A = \cos \delta_l. \tag{8.38}$$

Some care must be taken to include the correct multiple of π when using the inverse tangent function to obtain δ_l. A point of reference is given by the fact that the phase shift is zero when the potential is zero, which here translates to $a = 0$.

In the case of S-wave ($l = 0$) scattering, the results are particularly simple, since $j_0(kr) = \sin kr/kr$ and $n_0(kr) = -\cos kr/kr$. We find $\delta_0 = -ka$. The cross section then has the low-energy limit of

$$\frac{4\pi}{k^2} \sin^2 \delta_0 \longrightarrow \frac{4\pi}{k^2} (ka)^2 = 4\pi a^2. \tag{8.39}$$

This is four times the classical cross section for particles scattered from a sphere. The increase is a consequence of the wave nature of matter.

Soft Sphere

Nearly as straightforward is the case where the potential is a finite constant V_0 inside a radius a and zero outside. We will consider the situation where V_0 is positive and less than the energy E. The solution to the radial wave equation inside a is just $Arj_l(k'r)$, with $k' = \sqrt{2\mu(E - V_0)}/\hbar$. This form must match with $r \cos \delta_l j_l(kr) - r \sin \delta_l n_l(kr)$ at $r = a$.

The tangent of the phase shift can be extracted from the matching conditions. It is found to be

$$\tan \delta_l = \frac{k'j_l'(k'a)j_l(ka) - kj_k(k'a)j_l'(ka)}{k'j_l'(k'a)n_l(ka) - kj_k(k'a)n_l'(ka)}. \qquad (8.40)$$

For S-wave scattering, this reduces to

$$k \cot(ka + \delta_0) = k' \cot k'a \text{ or } \delta_0 = -ka + \cot^{-1}\left[\frac{k'}{k} \cot k'a\right]. \qquad (8.41)$$

In the high-energy limit, k'/k reduces to one, and δ_0 tends to zero. In the low-energy limit, k' becomes an imaginary constant, but we can still use these results. The term $(k'/k) \cot k'a$ tends to infinity in this limit; thus the hard-sphere result $\delta_0 \rightarrow -ka$ is reproduced.

8.3   *Computational Approach*

Partial wave analysis is done for a variety of potentials in the program **Scattr3D**. The program solves the radial equation for each desired angular momentum, extracts the phase shift, and constructs the associated contributions to the scattering wave function and the differential and total cross sections. The key step is the solution of the radial wave equation, which is done in procedure **SolveSchrodinger2** by a direct integration scheme[16] based on the Numerov algorithm.[17] The integration begins at $r = r_{\min}$ which may be zero, past $r_{\max}$ to r_{stop} in steps of size δr. Inside $r_{\min}$ the potential is assumed infinite and the wave function is forced to zero. The Numerov algorithm begins with this value of zero and with an arbitrary value assigned to u at $r_{\min} + \delta r$. The arbitrariness is removed when the function is normalized.

Once the integration is complete, the tangent of the phase shift is extracted by applying Eq. 8.20 at two selected radii r_1 and r_2. The correct normalization is also determined. The full wave function and the cross sections are computed from Eqs. 8.8, 8.29, and 8.30, with the sums restricted to a finite range selected by the user. The coefficient a_l that appears in Eq. 8.8 is related to the phase shift in Eq. 8.26.

The phase shift itself is not completely determined by the tangent. The correct multiple of π must be included. This multiple is fixed by comparing the number of nodes in the free and scattered radial wave functions. With each change in the difference between numbers of nodes there is added or subtracted a shift of π. With this convention, the phase shift vanishes at high energies, where the two wave functions are nearly identical.

The energy and length scales, V_0 and L_0, used in the program are user selectable, as is the reduced mass μ of the scattered particle. The scaled quantities are

$$\bar{r} = r/L_0, \quad \bar{E} = E/V_0, \quad \bar{V}(\bar{r}) = V(\bar{r}L_0)/V_0. \tag{8.42}$$

The radial wave equation Eq. 8.10 becomes

$$-\frac{d^2 u_{El}}{d\bar{r}^2} + \left[\frac{l(l+1)}{\bar{r}^2} + \zeta\bar{V}(\bar{r}) \right] u_{El} = \zeta\bar{E} u_{El}. \tag{8.43}$$

where $\zeta \equiv 2\mu V_0 L_0^2/\hbar^2$ is a dimensionless parameter equivalent to the parameter defined in chapter 3. When a mass is entered, the program computes a new value of ζ. The scaled wave number $\bar{k}$ can then be obtained from

$$\bar{k} \equiv kL_0 = \sqrt{2\mu L_0^2 E}/\hbar = \sqrt{\zeta\bar{E}}. \tag{8.44}$$

The units of the scales V_0 and L_0 and of the mass are also selectable within a set of preselected types. For mass the allowed units are MeV/c^2 and GeV/c^2; for energy they are eV, keV, MeV, and GeV; for length, they are nm, pm, and fm. Thus the program can be used for a wide range of scales, from atomic to nuclear.

8.4   Exercises

Below are exercises to guide your use of **Scattr3D**. Some require modification of the program. If a new potential is to be added, the necessary steps are discussed in comments at the beginning of the program unit **Sc3DPotl**. Before trying the exercises, review the section on running the program.

8.1   **Hard-Sphere Scattering**

Compare the numerical results for scattering from an impenetrable sphere with results obtained analytically in section 8.2.3. The hard-sphere potential is an explicitly listed choice under the **Potential** menu item.

8.2   **Square Well**

Compare numerical results for phase shifts with analytical results for a (spherical!) square well. This potential is an inversion of the soft sphere discussed in section 8.2.3, and can be selected in the program by choosing **Spherical shells** under the **Potential** menu item. A negative value for V_0 will produce a well.

8.3   **Attraction Versus Repulsion**

Compare the signs of phase shifts for attractive and repulsive Gaussian potentials. Selection of such a potential is available under the **Potential** menu item. The sign given to the value of the parameter V_0 determines the sign of the potential. Consider values of $|V_0|$ that are smaller and larger than the beam energy.

8.4   **Square Well Resonance**

Compute the first three phase shifts over a range of energies for a proton scattered from a square well of radius 4 fm and depth 50 MeV. Observe the behavior of the contributions of each partial wave to the total cross section, and determine the angular momentum of any resonance evidenced by a sharp peak in the cross section.

8.5 Free Particle Limit

Choose a potential, particle type, and beam energy, and verify that the scattering phase shifts tend to zero as the potential is reduced in strength.

8.6 Convergence in *l*

Compare the estimate in Eq. 8.32 for the maximum *l* required to what is actually needed to see convergence in the total cross section for scattering from a soft sphere. Is the estimated maximum sufficient for the differential cross section? The energy dependence should be observable; consider high and low energies, and wide and narrow targets. Study the form of the radial wave function for large *l* and explain why contributions from these partial waves are small.

8.7 Number of Bound States

Compare the number n_l of bound states of angular momentum *l* in a square well to the zero-energy limit of the phase shift δ_l. Use the hot key **Plot-Type** twice to cycle the display of the phase shift to a form that shows the full range of values. The program **Bound3D** may be used to compute the number of bound states. Levinson's theorem[5,18] states that the zero-energy limit is related by $\lim_{E\to 0}\delta_l = n_l\pi$.

8.8 Reduced Mass

Model the interaction between two neutrons as a square well of depth 35 MeV and radius 2 fm, and compute low-order phase shifts. Compare a correct calculation that uses the reduced mass and an incorrect one that does not. The entry of the reduced mass, or of the mass of any particle that is not one of the predefined types, can be accomplished by selecting the **Custom** type in the **Particle Type & Mass, Units** dialog box; this input box is obtained when **Particle Type & Mass, Units** is chosen under the **Parameters** menu item.

8.9 Yukawa Potential

Use the Yukawa potential $V(r) = V_0(L_0/r)\exp(-r/L_0)$, with $V_0 = -100$ MeV and $L_0 = 1.2$ fm, to model proton–proton scattering. What is the total cross section when the energy is 100 MeV in the center of mass frame?

8.10 Ramsauer–Townsend Effect

Determine the energies at which low-energy electrons experience no scattering from rare-gas atoms.[16] Use the models suggested in Exercise 8.7; however, keep in mind that the one-dimensional width corresponds to the three-dimensional diameter, not the radius. Compare results with those of the one-dimensional analysis and with data.[19]

8.11   Angular Momentum of a Resonance

Determine the angular momentum of each resonance[*][16,20] found when neutrons are scattered from a heavy element such as $^{56}_{26}$Fe. Use the predefined Woods–Saxon potential, $V(r) = V_0/[1 + \exp\{(r - c)/a\}]$, with $V_0 = 50$ MeV, $c = 1.17$ fm $\times A^{1/3}$, and $a = 0.75$ fm.

8.12   Differential Cross Sections

Characterize the differential cross sections for neutrons scattered from nuclei of different sizes. Use the Woods–Saxon potential to model the interaction, as described in Exercise 8.11. Under what conditions and at what energies does the differential cross section have the most structure?

8.13   Scattering Length

The low-energy behavior of phase shifts is such that $\tan \delta_l(k) \sim A_l k^{2l+1}$. This implies that zero-energy scattering is entirely S-wave, and that the total cross section is approximated by $\sigma \sim 4\pi A_0^2$. The negative of A_0 is called the scattering length a. More generally, one finds the expansion

$$k \cot \delta_0(k) = -\frac{1}{a} + \frac{1}{2} r_0 k^2 + \mathcal{O}(k^4),$$

where r_0 is known as the effective range. The two parameters a and r_0 can be used to fit much of low-energy scattering. Extract values for these parameters numerically, by recording S-wave phase shifts at a sequence of energy values and fitting the form of $\cot \delta_0(k)$, in the following cases:

a. Hard sphere of radius 2 fm.

b. Square well (an inverted soft sphere) of radius 2 fm and depth 35 MeV.

c. Yukawa potential, $V = V_0(L_0/r) \exp(-r/L_0)$, with $V_0 = -100$ MeV and $L_0 = 1.2$ fm.

8.14   Virtual Bound States

For a particle in a square well, or another potential, the centrifugal term acts as a barrier to the decay of states with energies and angular momentum greater than zero. These appear in scattering processes as resonances. The energy and wave function of such a virtual bound state can be estimated as the energy and wave function of a real bound state in an effective potential which is constant beyond the edge of the well, rather than proportional to $1/r^2$.

a. Use the program **Bound3D** to find a combination of square-well size and angular momentum that produces at least one such virtual bound state. Then verify that a resonance with the chosen angular momentum exists near this energy.

b. Compare the number of nodes in the bound state and scattering wave functions inside the well.

---

*For exercises that deal with resonances in one-dimensional potentials, see Exercises 8.8 and 8.9.

c. Increase the depth of the well until one resonance disappears and, finally, another appears. Compare this process with the changes in energy levels for (virtual) bound states in the square well.

d. Study the width of the resonance as a function of well depth, energy and angular momentum.[16,21]

8.15 Inverse-Square Potential

Compare numerical results for phase shifts associated with the potential $V(r) = V_0(L_0/r)^2$ with analytic results. The inverse-square potential is a specific form of power-law potential selectable under the **Potential** menu item. The analytic solution of the radial wave equation can be found by rearranging it into the form of Bessel's equation:

$$\left[\frac{d^2}{dx^2} + \frac{1}{x}\frac{d}{dx} + 1 - \frac{\nu^2}{x^2}\right]J_\nu(x) = 0.$$

The phase shift can be extracted by using the asymptotic form of the Bessel function J_ν for large x:

$$J_\nu(x) \sim \frac{\cos(x - \nu\pi/2 - \pi/4)}{\sqrt{x}}.$$

Projects

8.16 Electron Scattering

The scattering of electrons from an atom can be modeled by using the Lenz–Jensen potential[11,23]:

$$V(r) = -\frac{Ze^2}{r}e^{-x}(1 + x + b_2x^2 + b_3x^3 + b_4x^4),$$

where $x = 14.356 \, Z^{1/6}(r/\text{nm})^{1/2}$, $b_2 = 0.3344$, $b_3 = 0.0485$, and $b_4 = 2.647 \times 10^{-3}$. Add this potential to the program, and search for the resonance near 5 eV for $Z = 59$.

8.17 Absorption

In many experiments, beam particles are not only scattered, but also absorbed. The partial wave analysis can be modified to take this into account.[8] The partial wave amplitudes $f_l = [\exp\{2i\delta_l(k)\} - 1]/(2ik)$ are written as $f_l = [\eta_l(k)\exp\{2i\delta_l(k)\} - 1]/(2ik)$, with $|\eta_l| \le 1$. If $|\eta_l|$ is strictly less than one, flux will be removed from the outgoing partial wave. The values of η_l are parameters for a model of some process that results in this reduced flux. The cross section computed from the new f_l is then only the elastic cross section. The total cross section is obtained from the elastic and inelastic contributions. Given the following simple model for the η_l

$$\eta_l = \begin{cases} \frac{1}{2} & , l \le l_{\max} \\ 0 & , l > l_{\max}, \end{cases}$$

compute the elastic cross section for the scattering of neutrons from a square well of depth 35 MeV and radius 2 fm.

8.18   Transfer-Matrix Algorithm

A different algorithm can be used for the solution of the radial equation, one based on the transfer matrix approach.[25,26] This method allows an alternative approach to the determination of the branch of the inverse tangent that defines the phase shift. Add the transfer-matrix algorithm to **Scattr3D**. The program already contains the necessary structure to handle selection of an algorithm and contains a dummy procedure **Solve-Schrodinger1** that can be expanded. The program **Scattr1D** does include the transfer-matrix algorithm and can be used as an example, although the analysis in the present case requires use of Bessel functions rather than trigonometric functions.

8.19   Coulomb Phase Shift

When a $1/r$ potential is present, the phase shift analysis and the program must be modified because the spherical Bessel functions are no longer appropriate for the radial wave function at large distances. Instead one must use Coulomb waves, which are hypergeometric functions. A discussion of the solutions and the modified partial wave analysis is given by Messiah.[27] Add to **Scattr3D** a correct treatment of partial waves in the Coulomb potential.

8.5   Details of the Program

8.5.1   Running the Program

The menu options available for control of the program are as follows:

- **File:** - Get program information; read and write data files; exit the program.

  - **About CUPS:** Show description of software consortium.

  - **About Program:** Show credits and a brief description.

  - **Configuration:** Verify and/or change program configuration.

  - **New:** Set file name to default file name and start new calculation.

  - **Open . . . :** Open file and read contents.

  - **Save:** Save current state of the program to a file.

  - **Save As . . . :** Save current state of the program to a file with chosen name and set file name to this choice.

  - **Exit Program**

- **Parameters:** Set parameters.

– **Particle Type & Mass, Units:** Select particle type or a specified mass, energy unit, and length unit.

– **Numerical Parameters, Algorithm:** Change values of numerical parameters; select different algorithm, once others are installed.

● **Potential:** Display, choose, and modify the potential energy.

– **Display & Modify Current Choice:** Plot current choice for potential and display parameter values; allow modification.

– **Choose & Modify:**   Power-Law

–   Yukawa

–   Gaussian

–   Lennard-Jones

–   Woods-Saxon

–   Spherical Shells

–   Hard Sphere

–   User Defined

Choose this potential; plot and display parameter values; allow modification.

● **Compute:** Plot previous results, if any; then compute and plot new results. Except for the three-dimensional wave function, previous results are retained, in a different color, until a plot option is changed or a cleared plot requested.

– **Radial Wave Functions**

– **3D Wave Function:** Surface or contour plot of partial wave sum.

– **Phase Shifts:** Phase shifts as functions of energy.

– **Differential Cross Sections:** Partial wave sum versus angle.

– **Total Cross Sections:** Partial wave sum versus energy.

● **Help:** Display help screens.

– **Summary:** Display summary of menu choices.

– **'File':** Describe entries under **File**.

– **'Parameters':** Describe entries under **Parameters**.

– **'Potential':** Describe entries under **Potential**.

– **'Compute':** Describe entries under **Compute**.

– **'Algorithms':** Describe algorithms used in program.

Selection of options under **Parameters**, **Potential**, or **Compute** will bring up input screens for parameter values. Once values are entered for **Potential** or

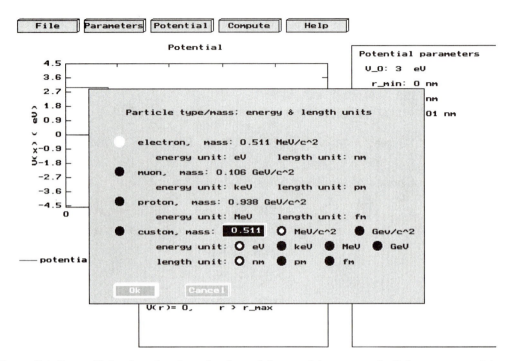

Figure 8.1: Input dialog box for the selection of the particle type and of the energy and length units.

Compute options, the new form is computed and plotted. Certain keys, especially some function keys, can then be used to alter the display or begin a new computation. Auxilliary input screens may appear, such as a screen for selection of a new plot type.

8.5.2   Sample Input and Output

Figure 8.1 shows the dialog box used for the selection of the particle type and the energy and length units used by the program. An example of a plot for a chosen potential is shown in Figure 8.2. Typical output of the program is shown in Figures 8.3–8.7. The option to show the three-dimensional wave function in the form of a contour plot is available, as are options for ranges of the phase shift and logarithms of the cross sections.

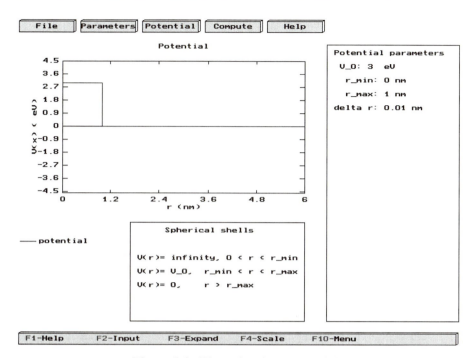

Figure 8.2: The soft sphere potential.

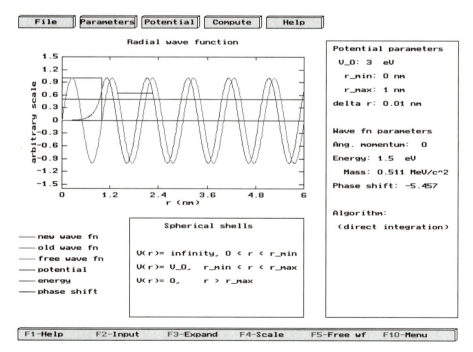

Figure 8.3: The radial wave function of an electron scattered from a soft sphere and the corresponding unshifted, free wave function.

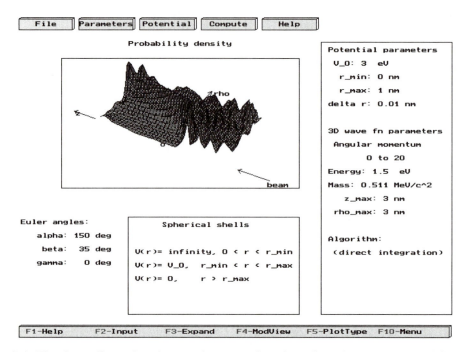

Figure 8.4: The three-dimensional scattering wave function of an electron scattered from a soft sphere. The number of partial waves included is 21.

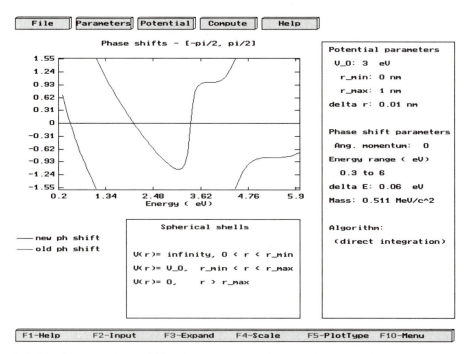

Figure 8.5: The S-wave phase shift of an electron scattered from a soft sphere. No old phase shift is shown; however, this option can allow comparison of two or more calculations done in succession.

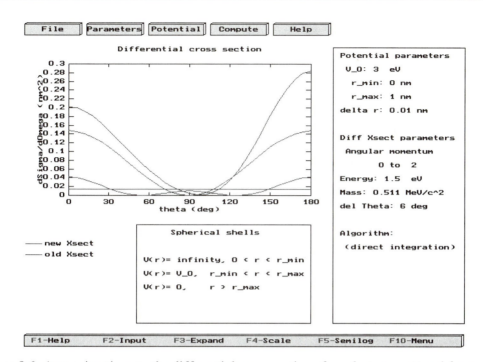

Figure 8.6: Approximations to the differential cross section of an electron scattered from a soft sphere. The topmost is the sum of the first three partial waves; the other three are the contributions of the individual partial waves.

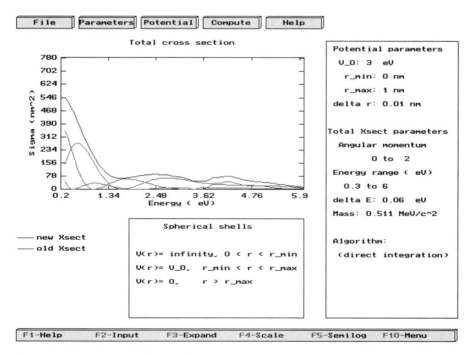

Figure 8.7: Approximations to the total cross section of an electron scattered from a soft sphere. The first three partial waves are shown along with their sum.

References

1. Prescott, C. Y., *et al.* "Parity non-conservation in inelastic electron scattering," Physics Letters **77B**(3):347, 1978. "Further measurements of parity non-conservation in inelastic electron scattering," *ibid.* **84B**(4):524, 1979.

2. Napolitano, J., *et al.* "Measurement of the differential cross section for the reaction $^2H(\gamma, p)n$ at high photon energies and $\theta_{c.m.} = 90°$," Physical Review Letters **61**(22):2530, 1988.

3. Bodek, A., *et al.* "Comparisons of deep-inelastic $e-p$ and $e-n$ cross sections," Physical Review Letters **30**(21):1087, 1973.

   Poucher, J. S., *et al.* "High-energy single-arm inelastic $e-p$ and $e-d$ scattering at 6 and 10°," *ibid.,* **32**(3):118, 1974.

4. Taylor, J. R. *Scattering Theory: The Quantum Theory on Nonrelativistic Collisions.* New York: Wiley, 1972.

5. Newton, R. G. *Scattering Theory of Waves and Particles.* New York: McGraw-Hill, 1966.

6. Merzbacher, E. *Quantum Mechanics,* 2nd ed. New York: Wiley, 1970, p. 231.

7. Anderson, E. E. *Modern Physics and Quantum Mechanics.* Philadelphia: Saunders, 1971, p. 396.

8. Gasiorowicz, S. *Quantum Physics.* New York: Wiley, 1974, p. 379.

9. Brandt, S., Dahmen, H. D. *The Picture Book of Quantum Mechanics.* New York: Wiley, 1985, p. 241.

10. Schmid, E.W., Spitz, G., Lösch, W. *Theoretical Physics on the Personal Computer.* Berlin: Springer-Verlag, 1988, p. 195.

11. Koonin, S. E., Meredith, D. C. *Computational Physics (FORTRAN Version).* Redwood City, CA: Addison-Wesley, 1990, p. 103.

12. Liboff, R. L. *Introductory Quantum Mechanics*, 2nd ed. San Francisco: Holden-Day, 1992, p. 724.

13. Park, D. A. *Introduction to the Quantum Theory*, 3rd ed. New York: McGraw-Hill, 1992, p. 291.

14. Liboff, R. L. *op. cit.,* p. 409.

15. Abramowitz, M., Stegun, I. A. (eds.) *Handbook of Mathematical Functions.* New York: Dover, 1964.

16. Greenhow, R. C., Matthew, J. A. D. "Continuum computer solutions of the Schrödinger equation," American Journal of Physics **60**(7):655, 1992.

17. Friar, J. L. "A note on the roundoff error in the numerov algorithm," Journal of Computational Physics **28**(3):426, 1978. Blatt, J. M. Practical points concerning the solution of the Schrödinger equation. *ibid.* **1**(3):382, 1967.

18. Levinson, N. "On the uniqueness of the potential in a Schrödinger equation for a given asymptotic phase," *Kgl. Danske Videnskab. Selskab, Mat.-fys. Medd.* **25**(9), 1949.

19. Massey, H. S. W., Burhop, E. H. S. *Electronic and Ionic Impact Phenomena.* London: Oxford University Press, 1952, p. 9.

20. Greenhow, R. C. "Computer modeling of resonance scattering in the time domain," American Journal of Physics **62**(3):240, 1994.

21. Greenhow, R. C., Matthew, J. A. D., Clark, R. M., Gates, G. A. "Relationship between the positions and breadths of single channel resonances," Journal of Physics. B (Atomic Molecular and Optical Physics) **24**(22):4677, 1991.

22. Koonin, S. E., Meredith, D. C. op. cit. p. 101.

23. Kieffer, L. J. "Low-energy electron-collision cross-section data III," *Atomic Data* **2**(4):293, 1971. Experimental data for comparison is given.

24. Gasiorowicz, S. *op. cit.* p. 384.

25. Kalotas, T. M., Lee, A. R., Howard, V. E. "Exact partial wave analysis for scattering by a segmented potential," American Journal of Physics **59**(3):225, 1991.

26. Gordon, R. G. "New method for constructing wave functions for bound states and scattering," Journal of Chemical Physics **51**(1):14, 1969. "Quantum scattering using piecewise analytic solutions," Methods in Computational Physics **10**:81, 1971. These contain careful discussion of improvements for accuracy, particularly for a linear approximation to the potential, but do not explicitly consider three-dimensional cases. See also the caveat in Senn, P. "Numerical solutions of the Schrödinger equation," American Journal of Physics **60**(9):776, 1992.

27. Messiah, A. *Quantum Mechanics*, vol. I. New York: Wiley, 1966, p. 421.

9

Bound States in Cylindrically Symmetric Potentials

John R. Hiller

9.1   Introduction

Although many interesting situations can be analyzed with the use of a one-dimensional Schrödinger equation, this is not always the case. Here we discuss some problems which can be reduced only to two-dimensional calculations. These include the quadratic Zeeman problem,[1] where a hydrogenic atom is placed in a uniform magnetic field, and the binding of the H_2^+ ion.[2-4] Each potential is cylindrically symmetric and is invariant under reflection through a suitably chosen origin on the axis of symmetry. In the Zeeman problem, the direction of the magnetic field specifies the symmetry axis, and for the H_2^+ ion, it is the line joining the two protons that specifies the axis. The Schrödinger equation is then reducible to a two-dimensional form that need be solved in only a quadrant.

To solve this equation, we use the method of evolution in imaginary time. This method exploits the connection between the time-dependent Schrödinger equation and a diffusion equation. The diffusion equation can be obtained from the time-dependent equation by substitution of an imaginary time; hence the name of the method. In the solution to the diffusion equation, which is obtained numerically, the higher-energy states decay most rapidly and, after a sufficient amount of "time," only the lowest state remains. One can view this as a minimization process where the expectation value for the Hamiltonian decreases with time until it reaches a minimum that is the eigenenergy of the lowest state.

The program **CylSym** was constructed to carry out this minimization/diffusion process. The user may select the potential to be considered and an initial guess for the wave function. Each contains adjustable parameters. In particular, the eigenvalues of parity and L_z, the component of angular momentum along the cylinder axis, may be chosen. The numerical solution of the diffusion equation is then carried out under user control. The form of the wave function and the expectation value of the Hamiltonian may be monitored. On convergence to the minimum, the eigenfunction and eigenenergy are obtained. In this way, systematic studies of the effects of changes in potential parameters can be carried out. For example, one can observe the effect on a hydrogenic ion of increases in the strength of the magnetic field to which it is exposed.

The remaining sections describe the physical situations, the numerical methods, and the program interface more fully.

9.2 Cylindrically Symmetric Potentials

9.2.1 General Discussion

We wish to consider the time-independent Schrödinger equation

$$-\frac{\hbar^2}{2\mu}\nabla^2\psi + V(\mathbf{r})\psi = E\psi \tag{9.1}$$

in situations where the potential V is cylindrically symmetric. The natural coordinates to use are, of course, cylindrical coordinates (ρ, ϕ, z), with the z-axis along the axis of symmetry. The potential is then a function of only ρ and z. The wave function may be chosen to be an eigenstate of $L_z = -i\hbar(\partial/\partial\phi)$, with eigenvalue $m\hbar$, and may be written in a factorized form:

$$\psi = \frac{1}{\sqrt{2\pi}}e^{im\phi}u_m(\rho, z), \quad m = 0, \pm1, \pm2, \dots \tag{9.2}$$

The Schrödinger equation then reduces to

$$Hu \equiv -\frac{\hbar^2}{2\mu}\left[\frac{\partial^2}{\partial\rho^2} + \frac{1}{\rho}\frac{\partial}{\partial\rho} - \frac{m^2}{\rho^2} + \frac{\partial^2}{\partial z^2}\right]u_m + Vu_m = Eu_m. \tag{9.3}$$

To reduce further the size of the numerical calculations, by a factor of two, we consider only potentials that are invariant under the reflection $z \to -z$. The potential then has the functional dependence $V(\rho, |z|)$ and is invariant under parity P, which takes $\vec{r}$ to $-\vec{r}$. The wave function may then be chosen to be an even or odd eigenstate of P:

$$P\psi = \pm\psi. \tag{9.4}$$

This implies that the factorized form obeys

$$P\frac{1}{2\pi}e^{im\phi}u_m(\rho, z) = \frac{1}{2\pi}e^{im(\phi+\pi)}u_m(\rho, -z) = \pm\frac{1}{2\pi}e^{im\phi}u_m(\rho, z). \tag{9.5}$$

Since $e^{im\pi} = (-1)^m$, u_m should be chosen to be either odd or even in z. Thus u_m need be calculated only in the quadrant where $0 \le \rho < \infty$ and $0 \le z < \infty$.

The boundary conditions for this quadrant are straightforward. Because we seek bound states, u_m must decrease to zero as $\rho \to \infty$ or $z \to \infty$. Along $z = 0$, u_m must be zero when odd in z and $\partial u_m/\partial z$ must be zero when u_m is even. Along $\rho = 0$, u_m must be zero when $m > 0$ and may be a non-zero constant when $m = 0$; these conditions follow from Eq. 9.3 and the requirement of normalizability.

The numerical method used is designed to find the lowest-energy bound state for a given pair of quantum numbers for L_z and P. These quantum numbers determine the specific form of the cylindrical wave equation (Eq. 9.3), where m is a parameter, and the reflection symmetry of u_m, which sets the choice of boundary conditions.

Given energy and length scales V_0 and L_0, we define useful scaled quantities

$$\bar{\rho} = \rho/L_0, \quad \bar{z} = z/L_0, \quad \bar{E} = E/V_0, \quad \overline{V}(\bar{\rho},\bar{z}) = V(\rho,z)/V_0. \quad (9.6)$$

The cylindrical wave equation (Eq. 9.3) then can be written in the form

$$-\left[\frac{\partial^2}{\partial\bar{\rho}^2} + \frac{1}{\bar{\rho}}\frac{\partial}{\partial\bar{\rho}} - \frac{m^2}{\bar{\rho}^2} + \frac{\partial^2}{\partial\bar{z}^2}\right]u_m + \zeta\overline{V}u_m = \zeta\bar{E}u_m. \quad (9.7)$$

where $\zeta \equiv 2\mu V_0 L_0^2/\hbar^2$ is, as in chapter 3, a dimensionless parameter. The program uses these scaled variables internally. The user selects the mass and potential parameters from which V_0, L_0, and ζ are computed.

Five situations that can be analyzed in this way are discussed in the next section. Two of them, the cylindrical well and the anisotropic oscillator, are such that the cylindrical equation can be separated into two independent equations and solutions found more directly. They are included, however, to provide a means to evaluate the accuracy of the numerical method.

9.2.2   Examples

Cylindrical Well

One of the simplest cylindrically symmetric potentials is a piecewise-constant potential that is negative inside a cylinder and zero outside. We write the functional form as

$$V = \begin{cases} -V_0 & , \rho < \rho_1, |z| < z_1 \\ 0 & , \text{otherwise}. \end{cases} \quad (9.8)$$

The parameter ρ_1 is then the radius of the cylinder, and $2z_1$ is the length. The natural energy scale is the depth V_0. For the length scale L_0, the program uses $\sqrt{\rho_1^2 + z_1^2}$, with the exception of the case where $\sqrt{\rho_1^2 + z_1^2}$ is zero. In that special case, L_0 is set to a single multiple of the chosen length unit.

Anisotropic Oscillator

The one-dimensional oscillator potential $\frac{1}{2}kx^2$, where the spring constant k is related to the angular frequency ω by $k = \mu\omega^2$, is easily generalized to a three-dimensional form: $\frac{1}{2}k_1x^2 + \frac{1}{2}k_1y^2 + \frac{1}{2}k_2z^2$. Notice that this form is not isotropic, since the spring constants k_1 and k_2 can be different. In cylindrical coordinates this potential is

$$V(\rho, z) = \frac{1}{2}\mu\omega_1\rho^2 + \frac{1}{2}\mu\omega_2z^2, \quad (9.9)$$

where $\omega_i = \sqrt{\mu/k_i}$. The eigenvalues are known to be

$$E_{n_1,n_2} = \hbar\omega_1(2n_1 + 1) + \hbar\omega_2\left(n_2 + \frac{1}{2}\right), \quad n_1, n_2 = 0, 1, 2, \ldots \quad (9.10)$$

Note, however, that there is some degeneracy.

For the purposes of the program, we define $V_0 = \hbar\omega_1$ as the energy scale and $L_0 = \sqrt{\hbar/\mu\omega_1}$ as the length scale. In terms of these parameters, the potential and eigenenergies are

$$V(\rho, z) = \frac{V_0}{2L_0^2}\left(\rho^2 + \left(\frac{\omega_2}{\omega_1}\right)^2 z^2\right) \quad (9.11)$$

and

$$E_{n_1,n_2} = V_0\left[2n_1 + \frac{\omega_2}{\omega_1}n_2 + 1 + \frac{\omega_2}{2\omega_1}\right]. \quad (9.12)$$

Quadratic Zeeman Effect

The Hamiltonian for a hydrogenic ion in a magnetic field is

$$H = -\frac{\hbar^2}{2\mu}\nabla^2 - \frac{Ze^2}{r} - \frac{e}{2\mu}\mathbf{B} \cdot (\mathbf{L} + 2\mathbf{S}) + \frac{e^2}{2\mu}\mathbf{A}^2, \quad (9.13)$$

where Ze is the nuclear charge, $\mathbf{B}$ the magnetic field, and $\mathbf{A}$ the vector potential. For a uniform field, we can choose $\mathbf{A} = \frac{1}{2}\mathbf{B} \times \mathbf{r}$. We also choose $\hat{z}$ along the direction of the field. The Hamiltonian then reduces to

$$H = -\frac{\hbar^2}{2\mu}\nabla^2 - \frac{Ze^2}{r} - \frac{eB}{2\mu}(L_z + 2S_z) + \frac{e^2B^2}{8\mu}\rho^2. \quad (9.14)$$

Thus there is a term quadratic in B, and it is the presence of this term that is of interest here.

The next-to-last term is the linear Zeeman term, which is usually studied in the context of perturbation theory. Since that analysis begins with a basis formed of the Coulomb eigenfunctions, the linear Zeeman term causes a mixing (as does the quadratic term). Here the energy eigenfunctions are chosen in such a way that they are already eigenfunctions of the linear term. Therefore, it adds only an m-dependent constant to the energy, and, for our purposes, can be ignored.

The potential that we use is then

$$V(\rho, z) = -\frac{Ze^2}{\sqrt{\rho^2 + z^2}} + \frac{e^2B^2}{8\mu}\rho^2. \quad (9.15)$$

Except for the case of very strong fields, the natural energy and length scales of the problem are the atomic ones of $V_0 = \alpha^2 mc^2$ and the Bohr radius $L_0 = \hbar c/\alpha\mu$, where α is the fine structure constant. In terms of these, we have

$$V(\rho, z) = V_0\left[-\frac{ZL_0}{\sqrt{\rho^2 + z^2}} + \frac{B^2}{8\mu^4\alpha^3}\left(\frac{\rho}{L_0}\right)^2\right]. \quad (9.16)$$

H_2^+ Ion

The binding of two nuclei into a molecule by exchange of a single electron[*] can be analyzed in the following approximate way. The light mass of the electron allows one to assume that the wave function of the electron adjusts instantaneously to any change in position of the slow-moving nuclei.[†] The nuclei can then be treated as fixed at some separation R, and the electron eigenvalue problem is solved for fixed centers of attraction. The total potential energy of the molecule is then approximated by the sum of the direct interaction $Z_1 Z_2 e^2 / R$ and the electron eigenenergy. Both contributions vary with R. If there is a region of negative potential energy with sufficient depth and width, the molecule will be bound.

The potential energy of the electron,

$$V(\mathbf{r}) = -\frac{Z_1 e^2}{|\mathbf{r} - \mathbf{R}_1|} - \frac{Z_2 e^2}{|\mathbf{r} - \mathbf{R}_2|}, \qquad (9.17)$$

is cylindrically symmetric with respect to an axis passing through $\mathbf{R}_1$ and $\mathbf{R}_2$. In cylindrical coordinates about this axis we may choose

$$\mathbf{R}_1 = \frac{1}{2} R\hat{z}, \quad \mathbf{R}_2 = -\frac{1}{2} R\hat{z}, \qquad (9.18)$$

so that

$$V = V(\rho, z) = -\frac{Z_1 e^2}{\sqrt{\rho^2 + (z - \frac{1}{2}R)^2}} - \frac{Z_2 e^2}{\sqrt{\rho^2 + (z + \frac{1}{2}R)^2}}. \qquad (9.19)$$

If the nuclear charges are equal ($Z_1 = Z_2 \equiv Z$), as they are in the case of H_2^+ ($Z_1 = Z_2 = 1$), then V is invariant under reflection in z.

The natural energy and length scales of the problem are $V_0 = \alpha^2 mc^2$ and the Bohr radius $L_0 = \hbar c/\alpha\mu$. In terms of these, we have

$$V = V_0 \left[-\frac{ZL_0}{\sqrt{\rho^2 + (z - \frac{1}{2}R)^2}} - \frac{ZL_0}{\sqrt{\rho^2 + (z + \frac{1}{2}R)^2}} \right]. \qquad (9.20)$$

Woods–Saxon Potential

The Woods–Saxon potential[5] is frequently used in nuclear shell models, where each nucleon is viewed as moving independently in an average central well. Although some nuclei are spherically symmetric, many are not, and thus the most general average potential is not spherically symmetric. We therefore include a distortion proportional to the Legendre polynomial $P_2(\cos \theta) = (3 \cos^2\theta - 1)/2$, where $\theta = \tan^{-1}(\rho/z)$ is the polar angle. This is the same type of distortion present in the anisotropic oscillator discussed above.

A Woods–Saxon potential distorted in this way is

$$V(r) = V_0 \frac{1 - bP_2(\cos \theta)}{1 + \exp[(r - c)/a]}. \qquad (9.21)$$

---

[*]For discussion of a model for such binding in one dimension, see Exercise 2.5.
[†]This is known as the Born-Oppenheimer approximation.

Clearly, V_0 sets the energy scale. Because c controls the width, it sets a natural length scale.

9.3 *Computational Approach*

9.3.1 Evolution in Imaginary Time

We replace the original elliptic eigenvalue problem given in Eq. 9.3 with a related parabolic diffusion problem[6]:

$$HU_m = -\hbar \frac{\partial}{\partial \tau} U_m,$$ (9.22)

where τ is the imaginary time.[*] A solution of the diffusion equation may be expressed as a sum over the eigenfunctions of the elliptic equation. Suppose that

$$Hu_m^{(n)} = E_n u_m^{(n)}, \quad n = 0, 1, 2, \ldots$$ (9.23)

Then

$$U_m(\rho, z, \tau) = \sum_{n=0}^{\infty} a_n e^{-E_n \tau/\hbar} u_m^{(n)}(\rho, z)$$ (9.24)

is readily shown to be a solution of Eq. 9.22. The coefficients a_n are determined by the initial shape $U_m(\rho, z, 0)$. At large times, the lowest-energy state dominates, provided a_0 is not zero. The function $u_m^{(0)}$ and energy E_0 can then be extracted from $U_m(\rho, z, \tau)$ if $U(\rho, z, 0)$ is chosen well and if computations are extended far enough in τ. One has

$$E_0 = \lim_{\tau \to \infty} \frac{\langle U|H|U \rangle}{\langle U|U \rangle}$$ (9.25)

and

$$u_m^{(0)} \propto \lim_{\tau \to \infty} e^{E_0 \tau/\hbar} U(\rho, z, \tau).$$ (9.26)

The approach taken is, then, to pick an initial guess for the wave function and to solve the diffusion equation for a time interval long enough to extract an accurate estimate for E_0. Convergence is most rapid if U at $\tau = 0$ is as close to $u^{(0)}$ as possible. If E_0 is negative, a constant S is added to the potential so that the exponential factor $\exp[-(E_0 + S)\tau/\hbar]$ does not grow with time. Otherwise, the calculation might be swamped by floating-point overflows in U_m. Since this shift in energy is a calculational device, it is automatically removed before an energy value is reported by the program.

9.3.2 Alternating Direction Implicit Method

The numerical method used for the diffusion equation is the alternating direction implicit (ADI) method.[7] It is a standard finite-difference method for parabolic

---

*Note that $\tau = it$ restores the time-dependent Schrödinger equation.

partial differential equations. Consider, for example, the two-dimensional diffusion equation

$$\frac{\partial^2 f}{\partial x^2} + \frac{\partial^2 f}{\partial y^2} = -\frac{\partial f}{\partial t}. \tag{9.27}$$

The method estimates the values of f at points (x_i, y_j) on a regular grid where $x_i = i\Delta x$ and $y_j = j\Delta y$ with Δx and Δy fixed spacings. The estimates are obtained only at equally spaced times $t_n = n\Delta t$. A discrete representation[7] of the differential equation can then be constructed.

A representation that leads to a simple explicit method is

$$\frac{f_{i+1,j}^n - 2f_{ij}^n + f_{i-1,j}^n}{\Delta x^2} + \frac{f_{i,j+1}^n - 2f_{ij}^n + f_{i,j-1}^n}{\Delta y^2} = -\frac{f_{ij}^{n+1} - f_{ij}^n}{\Delta t}, \tag{9.28}$$

where $f_{ij}^n \equiv f(x_i, y_j, t_n)$. It can be rearranged to give an explicit formula for f_{ij}^{n+1} in terms of f values at the previous time step. A more accurate method is an implicit one based on a different discrete representation

$$\frac{f_{i+1,j}^{n+1} - 2f_{ij}^{n+1} + f_{i-1,j}^{n+1}}{\Delta x^2} + \frac{f_{i,j+1}^{n+1} - 2f_{ij}^{n+1} + f_{i,j-1}^{n+1}}{\Delta y^2} = -\frac{f_{ij}^{n+1} - f_{ij}^n}{\Delta t}. \tag{9.29}$$

In this case a matrix problem must be solved to obtain the f_{ij}^{n+1}.

The ADI method is a compromise that retains the accuracy of an implicit method but results in tridiagonal matrix problems that are easy to solve quickly. To accomplish this, the representation alternates between two forms:

$$\frac{f_{i+1,j}^{n+1} - 2f_{ij}^{n+1} + f_{i-1,j}^{n+1}}{\Delta x^2} + \frac{f_{i,j+1}^n - 2f_{ij}^n + f_{i,j-1}^n}{\Delta y^2} = -\frac{f_{ij}^{n+1} - f_{ij}^n}{\Delta t},$$

$$\frac{f_{i+1,j}^{n+1} - 2f_{ij}^{n+1} + f_{i-1,j}^{n+1}}{\Delta x^2} + \frac{f_{i,j+1}^{n+2} - 2f_{ij}^{n+2} + f_{i,j-1}^{n+2}}{\Delta y^2} = -\frac{f_{ij}^{n+2} - f_{ij}^{n+1}}{\Delta t}. \tag{9.30}$$

These may be rearranged into linear matrix problems to be solved for all values of f at t_{n+1}, then t_{n+2}. The space direction along which the future values are used alternates, and the finite-difference formulas yield only an implicit solution; hence the name of the method. The rows and columns of the grid decouple to yield many smaller matrix problems, each of which is still tridiagonal. The resulting linear systems are solved by Crout reduction.[7]

Convergence can be monitored by observing the change in $\langle H \rangle \equiv \langle U|H|U \rangle / \langle U|U \rangle$ from one even time step to the next. The program **CylSym** displays this change as **delta$\langle H \rangle$**. Calculations are halted automatically when **delta$\langle H \rangle$** divided by $\langle H \rangle$ becomes less than a specified tolerance. The tolerance is set in the dialog box presented when the item **Time Step and Limits** is selected from the **Parameters** menu.

9.3.3   Finite-Difference Representation

The finite-difference representation used for the cylindrical wave equation is based on a variational principle. We first approximate $\langle U|H|U \rangle$ and $\langle U|U \rangle$ by sums over a grid, and then require that $\langle U|H|U \rangle$ be a minimum subject to the constraint that $\langle U|U \rangle$ is held fixed. To simplify the notation, time indices will be suppressed.

The analytic form of this variational process yields the Schrödinger equation. In terms of the scaled variables we have

$$\langle U|U \rangle = L_0^3 \int \bar{\rho}\,d\bar{\rho}\,d\bar{z}\,U^2 \tag{9.31}$$

and

$$\langle U|H|U \rangle = \frac{V_0 L_0^3}{\zeta} \int \bar{\rho}\,d\bar{\rho}\,d\bar{z}\,U \left[-\frac{\partial^2}{\partial \bar{\rho}^2} - \frac{1}{\bar{\rho}}\frac{\partial}{\partial \bar{\rho}} - \frac{\partial^2}{\partial \bar{z}^2} + \frac{m^2}{\bar{\rho}^2} + \zeta\bar{V} \right] U. \tag{9.32}$$

Integration by parts yields

$$\langle U|H|U \rangle = \frac{V_0 L_0^3}{\zeta} \int \bar{\rho}\,d\bar{\rho}\,d\bar{z} \left[\left(\frac{\partial U}{\partial \bar{\rho}}\right)^2 + \left(\frac{\partial U}{\partial \bar{z}}\right)^2 + \frac{m^2}{\bar{\rho}^2}U^2 + \zeta\bar{V}U^2 \right]. \tag{9.33}$$

To minimize $\langle U|H|U \rangle$ subject to the constraint we introduce a Lagrange multiplier E and minimize $\langle U|H|U \rangle - E\langle U|U \rangle$. The variation of $\langle U|U \rangle$ is

$$\delta\langle U|U \rangle = L_0^3 \int \bar{\rho}\,d\bar{\rho}\,d\bar{z}(U + \delta U)^2 - \int \bar{\rho}\,d\bar{\rho}\,d\bar{z}\,U^2,$$

$$= 2L_0^3 \int \bar{\rho}\,d\bar{\rho}\,d\bar{z}\,\delta U\,U + \mathcal{O}(\delta U^2). \tag{9.34}$$

The variation of $\langle U|H|U \rangle$ is

$$\delta\langle U|H|U \rangle = \frac{2V_0 L_0^3}{\zeta} \int \bar{\rho}\,d\bar{\rho}\,d\bar{z}\,\delta U \left[-\frac{\partial^2 U}{\partial \bar{\rho}^2} - \frac{1}{\bar{\rho}}\frac{\partial U}{\partial \bar{\rho}} - \frac{\partial^2 U}{\partial \bar{z}^2} + \frac{m^2}{\bar{\rho}^2}U + \zeta\bar{V}U \right]$$

$$+ \mathcal{O}(\delta U^2). \tag{9.35}$$

Therefore, for arbitrary δU, a zero variation implies Eq. 9.7.

The discrete calculation is done at the bottom of an infinite cylindrical well. At the edges of the well, the wave function is zero. The parameters that control the size of the well are the radius ρ_{max} and the ends of the axis $\pm z_{max}$. Provided ρ_{max} and z_{max} are sufficiently large, the results of any calculation will be essentially independent of their values. Finite values are necessary, however, to have a finite grid.

The grid is chosen in such a way as to avoid the singularities in $m^2/\bar{\rho}^2$ and V that lie at $\bar{\rho} = 0$. We use $\bar{\rho}_j = (j - \frac{1}{2})\Delta\rho$ and $\bar{z}_i = i\Delta z$ as the coordinates of the grid points, with $j = 1, \ldots, N_\rho$ and $i = 0, \ldots, N_z$. The intervals are related to the grid size by

$$\bar{\rho}_{max} = \left(N_\rho - \frac{1}{2}\right)\Delta\rho, \quad \bar{z}_{max} = N_z\Delta z. \tag{9.36}$$

The values of U at the grid points are denoted by $U_{ji} = U(\bar{\rho}_j, \bar{z}_i)$ and, similarly $\bar{V}_{ji} = \bar{V}(\bar{\rho}_j, \bar{z}_i)$.

The integrals in $\langle U|U \rangle$ and $\langle U|H|U \rangle$ are approximated by sums over this grid. We use the rectangle rule[*] for both the $\bar{\rho}$ and $\bar{z}$ integrals. Derivatives are approximated by central-difference formulas[7]

$$\left(\frac{\partial U}{\partial \bar{\rho}}\right)_{j+\frac{1}{2},i} = \frac{U_{j+1,i} - U_{ji}}{\Delta\rho}, \quad \left(\frac{\partial U}{\partial \bar{z}}\right)_{j,i+\frac{1}{2}} = \frac{U_{j,i+1} - U_{ji}}{\Delta z}. \tag{9.37}$$

---

[*]The rectangle rule for a one-dimensional integral $\int_a^{a+h} f(x)\,dx$ is simply the use of the approximation $hf(a + h/2)$.

These yield

$$\langle U|U\rangle = L_0^3 \Delta\rho\Delta z \sum_{j=1}^{N_\rho-1} \overline{\rho}_j \left[2 \sum_{i=1}^{N_z-1} U_{ji}^2 + U_{j0}^2 \right] \tag{9.38}$$

and

$$\begin{aligned}
\langle U|H|U\rangle = \frac{V_0 L_0^3}{\zeta} \Delta\rho\Delta z \sum_{j=1}^{N_\rho-1} & \left\{ 2 \sum_{i=1}^{N_z-1} \left[\overline{\rho}_{j+\frac{1}{2}} \left(\frac{U_{j+1,i} - U_{ji}}{\Delta\rho} \right)^2 \right. \right. \\
& \left. + \overline{\rho}_j \left(\frac{m^2}{\overline{\rho}_j^2} U_{ji}^2 + \zeta \overline{V}_{ji} U_{ji}^2 + \left(\frac{U_{j,i+1} - U_{ji}}{\Delta z} \right)^2 \right) \right] \\
& \left. + \overline{\rho}_{j+\frac{1}{2}} \left(\frac{U_{j+1,0} - U_{j0}}{\Delta\rho} \right)^2 + \overline{\rho}_j \left(\frac{m^2}{\overline{\rho}_j^2} U_{j0}^2 + \zeta \overline{V}_{j0} U_{j0}^2 + 2 \left[\frac{U_{j,1} - U_{j,0}}{\Delta z} \right]^2 \right) \right\}.
\end{aligned} \tag{9.39}$$

We then use a discrete form of the minimization procedure

$$\frac{\partial}{\partial U_{kl}} [\langle U|H|U\rangle - E\langle U|U\rangle] = 0 \tag{9.40}$$

to obtain

$$\begin{aligned}
-\frac{U_{j+1,i} - 2U_{ji} + U_{j-1,i}}{\Delta\rho^2} - \frac{1}{\overline{\rho}_j} \frac{U_{j+1,i} - U_{j-1,i}}{2\Delta\rho} - \frac{U_{j,i+1} - 2U_{ji} + U_{j,i-1}}{\Delta z^2} + \frac{m^2}{\overline{\rho}_j^2} U_{ji} \\
+ \zeta \overline{V}_{ji} U_{ji} = \zeta \overline{E} U_{ji}. \tag{9.41}
\end{aligned}$$

The left-hand side then provides the desired finite-difference representation of *HU* to be used in the ADI method. It has a form that one might choose immediately for *HU*, but we have the added reassurance that it corresponds exactly to the representation chosen for the expectation value.

9.4   *Exercises*

The exercises below give some guidance in the exploration of the physics that can be studied with **CylSym**. They include projects that require modification of the program. Before trying the exercises, review the section on running the program. For those projects that require addition of a new potential, the necessary steps are discussed in comments at the beginning of the program unit **ClSmPotl**.

9.1 **Infinite Cylindrical Well**

Verify that the program produces the correct result for an infinite cylindrical well. Since the program does all calculations inside such a well, this potential is obtained when one uses a finite cylindrical well that is as large as the computational grid. To set this potential, select **Cylinder** under **Potential** and enter values for ρ_1 and z_1 of the desired size. (See Eq. 9.8 for the definitions of these parameters.) Although the depth V_0 must remain non-zero, in order to set the energy scale, it represents only a constant shift that is easily taken into account. Once the potential parameters have been entered, change the grid size, specified by $\rho_{\max}$ and $z_{\max}$, to match the chosen size of the well; the grid dialog box is available via the function keys and the **Parameters** menu.

9.2 Anisotropic Oscillator

Show that the program obtains the correct result for the anisotropic oscillator, which is discussed as the second example in section 9.2.2. Consider cases where the magnetic quantum number m is not zero. Consider both parities.

9.3 Finite Cylindrical Well

Compare results from **CylSym** for the finite cylindrical well, as defined in Eq. 9.8, with those obtained by first separating the eigenvalue problem into two one-dimensional problems and then solving each with use of **Bound1D**. The separation of variables will result in an ordinary square-well wave equation in the z-variable, but a slightly more complicated equation in the ρ-variable. The latter can be converted to a standard one-dimensional wave equation by factoring $1/\sqrt{\rho}$ from the wave function.

9.4 Perturbative Zeeman Effect

Compare numerical results for the quadratic Zeeman effect to first-order perturbation theory. (The potential is given in Eq. 9.15.) At what field strength do they begin to differ by 10%?

9.5 Variational Estimate for the Quadratic Zeeman Effect

Try simple variational wave functions such as $e^{-\alpha r}$ or $e^{-\beta r^2}$ for the state of the electron in the quadratic Zeeman problem, and compare the energies obtained with numerical results from the program.

9.6 Variational Estimate for the H_2^+ Ion

Repeat Exercise 9.5 for the H_2^+ ion, which is discussed as the fourth example in section 9.2.2. A simple guess for a variational wave function in this case is $e^{-\alpha r_1} + e^{-\alpha r_2}$, where $r_1 = |\mathbf{r} + \frac{1}{2}R\hat{z}|$ and $r_2 = |\mathbf{r} - \frac{1}{2}R\hat{z}|$ with R the nuclear separation.[8,9] For some of the integrals, a change of variables to r_1, r_2, and the azimuthal angle can be useful. (See also the discussion by Park.[9]) The best values of the variational parameters, such as α in the example, will vary with R. Once the dependence on R is determined, the effective potential $V_{eff}(R)$ for the two protons can be constructed and analyzed. At what separation is V_{eff} minimized? Compare this with the result obtained with the program.

9.7 Distorted Woods–Saxon Potential

Determine the effect of distortion on the eigenenergies of the Woods–Saxon potential specified by Eq. 9.21. Compare the energy shift to that found in perturbation theory.

9.8 Diamagnetic Susceptibility

Compute the diamagnetic susceptibility χ for hydrogenic ions by comparing the field strength dependence of the energy to $-\frac{1}{2}\chi B^2$. This requires computation of the expectation value of the quadratic Zeeman Hamiltonian, discussed in section 9.2.2, for a series of field strength values, and a fit to a quadratic polynomial in B.

9.9 H_2^+ Binding

a. Determine the effective potentials for two protons when an electron is in the lowest-energy parity-even and parity-odd states. As discussed in section 9.2.2, the effective potential is the sum of the direct interaction e^2/R and the electron eigenenergy. The program can compute the electron eigenenergy for a series of separations. If carefully chosen, this series will provide enough information for a useful plot of the effective potential. Which parity case is more likely to have a bound state for the protons?

b. Use the effective potential and the program **Bound3D** to look for bound states and determine the binding energy. Note that the binding energy of the ion should be determined relative to the energy of a hydrogen atom and a proton infinitely separated. Be certain to use the correct reduced mass.

Experiments done with H_2^+ find a binding energy of 2.8 eV; the minimum in the effective potential should be near 0.11 nm.

9.10   **Muomolecules**
Consider the binding of H_2^+ when the electron is replaced by a muon.[10] Compare the binding energy and size of the muonic ion to the electronic one. For some discussion of the necessary steps, see Exercise 9.9. The replacement of the electron by the muon in the program is accomplished within the dialog box obtained by the selection of **Particle Type & Mass, Units** under **Parameters**.

Projects

9.11   **Anharmonic Oscillator**
Add a term proportional to z^4 to the potential for the anisotropic oscillator. An entry for the coefficient should be added as a parameter for the new potential. Compare results with those obtained by perturbation theory and the variational method. A simple trial wave function is a product of harmonic-oscillator Gaussians in which the angular frequencies are treated as variational parameters.

9.12   **Stark Effect**
The potentials included can be modified to study the effect of a constant electric field on an anisotropic oscillator, a hydrogen atom, or a H_2^+ ion. Before any numerical calculations are done, determine the range of field strengths for which a bound state calculation makes sense. (The original atomic experiment was done with a strength of 10^5 V/cm.) The field may be generated by a pair of capacitor plates placed beyond the range of the calculational grid. Compare results with perturbation theory. In the case of the oscillator, compare also the exact solution. In first-order perturbation theory, the shift in the energy due to a field of strength $\mathscr{E}$ is $\frac{1}{2}\alpha\mathscr{E}^2$, where α is the polarizability. Use a fit to this form to extract an estimate for α.

9.13 Heavy Baryons

A heavy baryon consists of three quarks, one or more of which is heavy. If two are heavy and the third relatively light, one can consider a Born–Oppenheimer approximation to the three-body bound state. To be specific, consider a bbs system: two bottom quarks and a strange quark. The strange quark is much lighter than the other two, but not so light that the nonrelativistic Schrödinger equation cannot be used. The two bottom quarks are then bound together by the strange quark and their own direct interaction.

Estimate the mass of such a baryon by first using **CylSym** to compute the effective potential between the bottom quarks, and then using **Bound3D** to compute the binding energy. To use **CylSym** for the first step, add a potential that uses an appropriate model for the quark interaction. The Cornell model[11]—$V(r) = -g/r + a^2 r$—is a reasonable choice. The values of the parameters are $g = 0.52\ \hbar c$ and $a = 0.427\ \text{GeV}/\sqrt{\hbar c}$. The quark masses can be taken to be 500 MeV/c^2 for the strange quark, and 5 GeV/c^2 for the bottom quark.

9.14 HHe$^+$ Binding

Can a single electron bind a proton to an alpha particle? Does a muon do a better job? Use the program to answer these questions. The assumption that V is even in z must be abandoned and the code modified to use the full interval $[-z_{max}, z_{max}]$. The steps discussed in Exercise 9.9, with appropriate modifications, can then be used to determine the binding energies. In particular, the direct interaction is now $2e^2/R$.

9.15 Two-Dimensional Problems

The program can be modified to solve the Schrödinger equation in two Cartesian dimensions:

$$-\frac{\hbar^2}{2\mu}\left(\frac{\partial^2}{\partial x^2} + \frac{\partial^2}{\partial y^2}\right)\psi + V(x,y)\psi = E\psi.$$

Two simple potentials are the square well, perhaps combined with a perturbation proportional to xy, and the anisotropic oscillator, with an optional anharmonic term such as $xy(x^2 + y^2)$. Carry out the necessary modifications for one of these potentials and obtain the ground state energy and wave function.

9.16 Two-Body Problems

A two-body problem in one dimension has the same number of degrees of freedom as a one-body problem in two dimensions, and the Schrödinger equations take nearly the same form in the two cases. The program can therefore be modified to treat the two-body problem.* An interesting interaction is that of two oscillators coupled by a linear force. Another is that of two one-dimensional "atoms," modeled as two particles bound to two adjustable centers of attraction. Notice that if two coupled oscillators have the same frequency, the two-dimensional equation can be separated into

---

*The time evolution of two-body systems can be observed with the program **Ident**.

an analytically soluble pair of one-dimensional equations; this can provide a useful check on the results of the modified program. One can consider distinguishable particles, or fermions, or bosons. For identical particles, the interactions must be invariant under interchange of their coordinates x_1 and x_2; calculations may then be restricted to $x_1 \geq x_2$. Select one of these cases and modify the program accordingly. Use the new program to compute energies and wave functions. In the case of two atoms, investigate the possibility of binding by computing the effective potential between the centers of attraction as a function of their separation.

9.17 Excited States

The energies and wave functions of excited states can be computed by the method used in **CylSym** if the wave functions of the lower states are projected out. This of course requires that the wave functions of the lower states be calculated first. The projection $\mathcal{P}$ uses them explicitly to subtract their contribution to the sum Eq. 9.24 as follows:

$$\mathcal{P}U_m = U_m - \sum_{n=0}^{n'-1} u_m^{(n)}(\rho, z) \int \rho d\rho dz u_m^{(n)}(\rho, z)U_m(\rho, z, t),$$

where n' is the index of the excited state that is desired and normalization of the u_m is assumed. With the projection, the coefficients a_n in the eigenfunction expansion Eq. 9.24 are zero for E_n less than the energy of the desired excited state. Because of round-off errors in the ADI calculations, particularly on the odd steps, the projection must be done after each even-numbered step or after some small multiple of even steps. Round-off errors in the projections themselves limit the number of excited states that can be studied in this way. Modify the program to compute the first-excited even-parity state of an electron bound to two protons, and determine whether this state results in binding of the protons.

9.5   Details of the Program

9.5.1   Running the Program

The program is controlled by choosing menu options and functions keys. The menu options are as follows:

- **File:** - Get program information; read and write data files; exit the program.

  - **About CUPS:** Show description of software consortium.
  - **About Program:** Show credits and a brief description.
  - **Configuration:** Verify and/or change program configuration.
  - **New:** Set file name to default file name and start new calculation.
  - **Open ...:** Open file and read contents.
  - **Save:** Save current state of the program to a file.

– **Save As...:** Save current state of the program to a file with chosen name and set file name to this choice.

– **Exit Program**

● **Parameters:** Choose parameters that control the calculation and display of results.

  – **Particle Type & Mass, Units:** Select particle type and mass and units for energy and length.

  – **Grid Size and Spacings:** Select size and spacings used in calculation.

  – **Time Step and Limits:** Select time step, limit on number of steps, tolerance, and the artificial shift in the potential.

  – **Viewpoint:** Change view of surface plot from within a special input screen.

● **Potential:** Display, choose, and modify the potential energy.

  – **Display & Modify Current Choice:** Plot current choice for potential and display parameter values; allow modification.

  – **Choose & Modify:**   **Zeeman**

  –   **H2+ Ion**

  –   **Woods-Saxon**

  –   **Oscillator**

  –   **Cylinder**

  –   **User-Defined**

  Choose this potential; plot and display parameter values; allow modification.

● **Initial Wf:** Display, choose, and modify the initial guess for the wave function.

  – **Display & Modify Current Choice:** Plot current choice for initial wave function and display parameter values; allow modification.

  – **Choose & Modify:**   **Coulombic**

  –   **Two-Ctr Coulombic**—two-centered Coulombic wave function for ion.

  –   **Gaussian**

  –   **Bessel-Cosine**—product of J_m and cosine.

  –   **User-Defined**

  Choose this initial guess; plot and display parameter values; allow modification.

● **Help:** Display help screens.

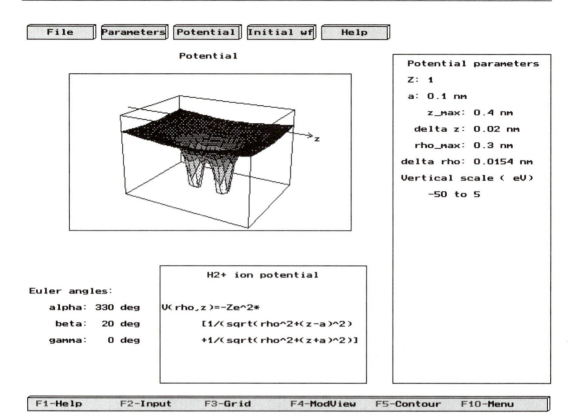

Figure 9.1: The potential seen by the electron in the H_2^+ ion.

- **Summary:** Display summary of menu choices and function keys.

- **'File':** Describe entries under **File**.

- **'Parameters':** Describe entries under **Parameters**.

- **'Potential':** Describe entries under **Potential**.

- **'Initial Wf':** Describe entries under **Initial Wf**.

- **How To Use:** Describe how to use the program.

The options under **Parameters, Potential**, and **Initial Wf** lead to input screens for the selection of relevant parameters.

The main set of function keys are defined as follows:

- **F1** Show help screen that briefly describes menu options and function keys.

- **F2** Begin time steps. While steps are in progress, **F2** can be used to halt the calculation. Results of the last complete step are retained.

- **F3** Take single time step.

- **F4** Undo last set of time steps. If a single step was taken, then only that step will be undone.

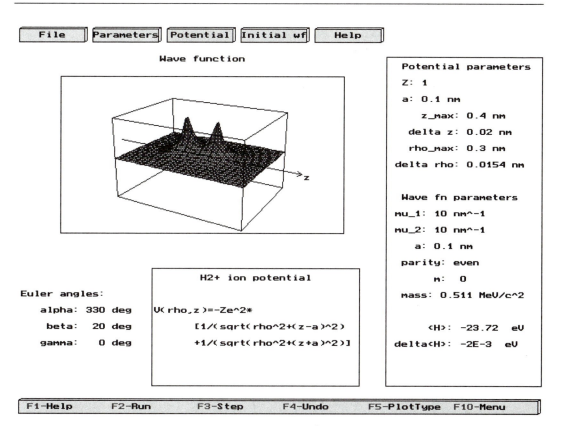

Figure 9.2: The wave function of the electron in the H_2^+ ion with the protons held at the selected separation.

- **F5** Change plot type. An input screen will appear with a list from which to make a selection.

- **F10** Activate menu. A menu selection can then be made using arrow keys and the <**Enter**> key. This is unnecessary if a mouse is used.

9.5.2 Sample Input and Output

The dialog box used for the selection of the particle type and the energy and length units is shown in Figure 9.1; the same dialog box is used by both **CylSym** and **Scattr3D**. An example of a plot for a chosen potential is shown in Figure 9.1. Typical output of the program is shown in Figures 9.2 and 9.3. The option to show the wave function or the potential in the form of a contour plot is available.

References

1. Anderson, E. E. *Modern Physics and Quantum Mechanics.* Philadelphia: Saunders, 1971, p. 349.

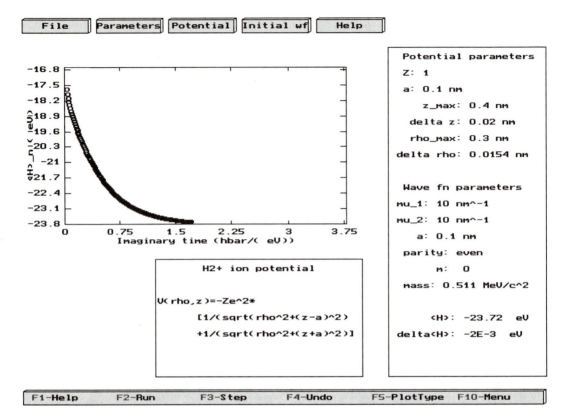

Figure 9.3: A history of the expectation value of the Hamiltonian as a function of imaginary time.

2. Gasiorowicz, S. *Quantum Physics*. New York: Wiley, 1974, p. 316.

3. Eisberg, R., Resnick, R. *Quantum Physics of Atoms, Molecules, Solids, Nuclei and Particles,* 2nd ed. New York: Wiley, 1985, p. 418.

4. Park, D. A. *Introduction to the Quantum Theory,* 3rd ed. New York: McGraw-Hill, 1992, p. 502.

5. Woods, R. D., Saxon, D. S. "Diffuse surface optical model for nucleon–nuclei scattering," Physical Review **95**(2):577, 1954. See also Preston, M. A., Bhaduri, R. K. *Structure of the Nucleus.* Reading, MA: Addison-Wesley, 1975, p. 97.

6. Koonin, S. E., Meredith, D. C. *Computational Physics (FORTRAN Version).* Redwood City, CA: Addison-Wesley, 1990, p. 181.

7. Gerald, C. F., Wheatley, P. O. *Applied Numerical Analysis,* 4th ed. Reading, MA: Addison-Wesley, 1989.

8. Davydov, A. S. *Quantum Mechanics.* Reading, MA: Addison-Wesley, 1965, p. 478.

9. Park, D. A. *op cit.,* p. 503.

10. Bracci, L., Fiorentini, G. "Mesic molecules and muon catalyzed fusion," Physics Reports **86**(4):169, 1982.

11. Eichten, E., Gottfried, K., Kinoshita, T., Lane, K. D., Yan, T.-M. "Charmonium: the model," Physical Review D **17**(11):3090, 1978; "Charmonium: comparison with experiment," *ibid.,* **21**(1):203, 1980.

Appendix A

Walk-Throughs for All Programs

These "walk-throughs" are intended to give you a quick overview of each program described in this book. Please see the Introduction for one-paragraph descriptions for all programs produced by the CUPS project.

A.1 Walk-Through for Bound1D *Program*

When the program is first run an information screen appears. Press **Enter** (or any visible character key) to get rid of this. You will then see a square potential well 400 eV deep and 0.2 nm wide (the default) plotted on the screen.

Finding Eigenvalues

This part of the program is used to find the eigenfunctions and eigenvalues of a number of different potential wells. For the purpose of this walk-through, it will be more informative to choose a well shape not so symmetric as the default.

- **Click on the menu item Potential**. This will offer you a choice of potential shapes. Select **Ramped Well**. This will change the potential to one which is linear within a range of 0.2 nm, and zero outside that range.

- **Select the menu item Parameters | Vary Well Parameters**. This allows you to vary the depth or the width of this well. Set the depth to 400 eV, and press **Enter** to accept that value.

- **Select the menu item Method | Try Energy (with mouse)**. Place the mouse within the graph somewhere near an energy of -330 eV ($E_B = 330$ eV). A (normalized) solution of the Schrödinger equation with that value of the energy will appear. It will have the "correct" behavior at the left-hand edge of the screen, but will be much too large at the right-hand edge. Move and click the mouse around this value, and see if you can find the energy for which the solution has the "correct" behavior at both edges.

- You should have found that there is an eigenvalue of this potential well somewhere between 300 and 320 eV. Return to the main menu with the **F 10** hot key, **select the menu item Method | Hunt for Zero**, and enter these two numbers when it asks you for input. The program will find the exact value by a binary chopping procedure. When it is finished, press **Enter** and it will display the corresponding eigenfunction, whose energy should be very close to 312 eV.

- **Select the menu item Spectrum | Find Eigenvalues**. The program will automatically find all the bound state eigenvalues. You may inspect any of the eigenfunctions by selecting **Spectrum | See Wave Functions**. When you have finished, return to the main menu by choosing **Cancel** or pressing **Esc**.

Wavefunction Properties

Select the menu item Parts | Part 2: Wave function Properties. This part of the program can be used to calculate various overlap integrals involving the eigenfunctions you found in part 1, or linear combinations of them.

- **Click on the menu item Psi 1** and select one of the quantum numbers shown. That eigenfunction will appear in the uppermost viewport.

- **Click on the menu item Psi 2** and select a different quantum number. That eigenfunction will appear in the second viewport, and the *product* of the two eigenfunctions you chose will appear in the third viewport.

- **Select the menu item Integrate**. The area under the product function will be calculated. Since you chose two different eigenfunctions they should be orthogonal to one another. Therefore the integral should vanish, to the accuracy of the calculation.

- **Select the menu item Operator | x**. The second viewport will now contain the product of x with the second eigenfunction. The product in the third viewport will also change accordingly. Selecting **Integrate** again will calculate the matrix element of x between the two eigenstates.

Time Development

Select the menu item Parts | Part 3: Time Development. This part of the program will show how linear combinations of the eigenfunctions change with time.

- **Select the menu item Wave Func | Choose Wave Function**. You will be asked to supply coefficients with which to construct a linear combination of eigenstates. The default values are 1.000 for $n = 1$ and zero for all others. Enter the value 0.7071 (i.e., $1/\sqrt{2}$) against $n = 1$ and $n = 2$. Press **Enter** and a wave function will appear which is composed equally of those two states. It is plotted so that the vertical axis measures the amplitude of the function, and the phase is represented by color according to the color wheel shown.

- **Select the menu item Wave Func I Show Time Development** and the wave function will be animated to show how it changes with time. Stop the animation and return to the main menu by pressing the hot key **F 10**.

- **Select the menu item Plot How I Real and Imaginary**. The wave function will be replotted with the real part in red and the imaginary part in green. Repeat the previous menu operations and see the time development of both parts separately.

- **Select the menu item Measure I Position**. The expectation value of x at the current time will be calculated and displayed.

A.2   *Walk-Through for* Scattr1D *Program*

The initial screen comes up with the default potential, a square barrier.

- **Select Display & Modify Current Choice from the Potential menu.** Use the resulting input screen to enter a new value of -2 eV for V_0. The minus sign flips the potential into a square well.

- **Select Trans & Refl Probabilities from the Compute menu.** In the input screen that appears, enter an energy range of 0.1 eV to 3.0 eV, with steps (delta E) of 0.02 eV to be taken. The plot that develops shows the transmission and reflection probabilities as functions of energy, and clearly indicates resonances.

- **Press ⟨F3⟩ to enlarge the plot.**

- **Press ⟨F4⟩ to rescale the plot.** The dialog box that is presented can be used to select a range of energies. Choose a range that eliminates all but one transmission peak, and read off the energy at the location of the peak. (The first is located at approximately 0.35 eV.) Repeat this procedure to obtain the location of one valley. (The first is at ~0.65 eV.)

- **Select Wave Function from the Compute menu.** Then enter the energy of the selected peak.

- **Press ⟨F4⟩ to change the type of plot.** Choose from any of the other three types.

- **Press ⟨F5⟩ to separate the incident wave from the reflected.**

- **Press ⟨F2⟩ to return to the energy input screen.** Try the value associated with the valley in transmission.

- **Select Exit Program from the File menu.** The program will not allow an immediate exit if calculations have been done but not saved. There is instead an opportunity to save the work to a file. This can be done directly with other entries in the **File** menu. Notice that work saved previously, perhaps in preparation for a lecture demonstration, can be recalled with the **Open...** entry.

A.3   Walk-Through for QMTime Program

Start the program, then choose **File | About Program** for a brief description of the display. (A thorough description is given in Appendix B.)

- **The free particle.** Choose **Run** to see the default wave function (Gaussian) evolve in time under the default potential (flat). Choose **File | Things to Notice** for some pointers about what you're seeing.

- **Momentum space wave functions.** Choose **Reset | Reset this Run**, then **Display | Plot What**, and turn on the display of momentum space wave functions. Take a few moments to guess how the momentum space wave function will change as time evolves, then choose **Run** to check your guess.

- **Scattering.** Choose **Potential | Step** (accept the defaults in this and all subsequent input screens), and then **Wave func | Gaussian**. Use **Run** to start the simulation.

- **Oscillating.** Use **Display | Plot What** to turn off the display of momentum space wave functions. Then choose **Potential | Harmonic Oscillator** and **Wave func | Combination of Energy States**. Choose **Run** to see the particle bounce back and forth in the harmonic oscillator potential well. After one period has passed (say, the wave packet moving from far left to far right and back again), does the wave function return to its original condition?

- **Coherent states.** Choose **Wave func | Combination of Energy States** and use the input screen to put the system into its ground state. Then choose **Wave func | Shift**. You have produced one of the so-called coherent states. What is special about the time development of such states? (Hint: Use **Display | Plot How** to show only the probability density.)

A.4   Walk-Through for Latce1D Program

When the program is first run an information screen appears. Press **Enter** (or any visible character key) to get rid of this. You will then see a "lattice" of six square potential wells (the default) plotted on the screen.

- **Click on the menu item Wells**. This will offer you a choice of potential shapes and number. Select **Number of Wells**. This will allow you to change the number of wells in the lattice. Set this number to 8.

- **Select the menu item Wells | Well Parameters**. This allows you to vary the depth or the width of this well. Set the depth to 400 eV.

- **Select the menu item Method | Try Energy (with mouse)**. Place the mouse within the graph somewhere near an energy of -350 eV ($E_B = 350$ eV). A solution of the Schrödinger equation with that value of the energy will appear. It will have the "correct" behavior at the left-hand edge of the screen, but will be much too large at the right-hand edge. Move and click the mouse around this value, and see if you can find the energy for which the solution has the "correct" behavior at both edges. This will prove to be extremely difficult to do. There are a lot of energy levels and they are very close together.

- **Select the menu item Method | Solve Range of Energies**. An input screen will appear asking you to supply three numbers—a lowest value for the binding energy, a highest value, and a step size. (Positive numbers are required.) Default values of 0, 400, 4 eV appear with the screen. Change these to 350, 370, 0.5. Then press **Enter**. The Schrödinger equation will be solved for this whole range of energies and the asymptote plotted against energy. You will see that the curve crosses the energy axis eight times. That means there are eight values of the energy for which the solution has the right asymptotic behavior to be an eigenvalue. Locate one of these values approximately, so that you can say definitely that it lies between two bounds. If you need to, you can re-enter values for the range by using the hot key **F2**.

- **Select the menu item Method | Hunt for Zero**, and enter these two numbers when it asks you for input. The program will find the exact value by a binary chopping procedure. When it is finished, press **Enter** and it will display the corresponding eigenfunction.

- **Select the menu item Spectrum | Find Eigenvalues**. The program will automatically find all the bound state eigenvalues. You should observe that there are 24, and the lowest 16 lie in two narrow bands. You may inspect any of the eigenfunctions by selecting **Spectrum | See Wave Functions**.

- **Select the menu item Spectrum | Sum Probabilities**. You will be asked to supply two level numbers. Accept the default values of 1 and 8. The program will then calculate and display the probability density for each of the levels 1–8 in order. It will also sum all these and display the result. You should observe this sum to be the same in the vicinity of all the wells of the lattice.

- **Select the menu item Lattice | Irregular Lattice**. Then choose **a regular lattice with one shallow well**. You should then see a lattice with an "impurity atom" in it. Repeat the previous two operations and observe that there is one state in what was previously the energy gap between the two lowest bands. It is level number 8. Observe how this state is firmly localized in the vicinity of the impurity. To see this use **Spectrum | Sum Probabilities** with $n = 1$–7 and $n = 8$ separately.

- **Select the menu item Lattice | Regular Lattice**. Then select **Lattice | Apply Electric Field**. Enter the value of 30 V. You should observe the potential to be what it was before with an extra linear component superimposed. Find all the energy levels as previously. Observe states 1 and 8 particularly.

An interpretation of what you see is that the electron in a state at the bottom of a band behaves like an ordinary electron (it moves toward the region of lowest potential). At the top of the band it behaves as though it has either a positive charge or a negative effective mass. Conversely, if you sum probabilities for levels 1–7, you will observe a "hole" at the edge of the lattice where the potential is most negative (i.e., the potential energy is highest).

A.5 *Walk-Through for* Bound3D *Program*

When the program is first run an information screen appears. Press **Enter** (or any visible character key) to get rid of this. You will then see a radial representation of a three-dimensional "square" potential well with depth 300 eV and radius 0.05 nm (the default) plotted on the screen.

Finding Eigenvalues

This part of the program is used to find the eigenfunctions and eigenvalues of a number of different potential wells.

- **Click on the menu item Potential**. This will offer you a choice of potential shapes, as well as the ability to change the parameters of any of them. Select **Vary Well Parameters**. Set the radius to 0.10 nm.

- **Select the menu item Method | Try Energy (with Mouse)**. Place the mouse within the graph somewhere near an energy of -270 eV ($E_B = 270$ eV). A solution of the Schrödinger equation with that value of the energy will appear. It will have the "correct" behavior at the left-hand edge of the screen—i.e., it will be zero at $r = 0$—but it will be much too large at the right-hand edge. Move and click the mouse around this value, and see if you can find the energy for which the solution has the "correct" behavior at both edges.

- You should have found that there is an eigenvalue of this potential well somewhere between 250 and 280 eV. **Select the menu item Method | Hunt for Zero**, and enter these two numbers when it asks you for input. The program will find the exact value by a binary chopping procedure. When it is finished, press **Enter** and it will display the corresponding radial eigenfunction.

- **Select the menu item Spectrum | Find Eigenvalues**. The program will automatically find all the bound state eigenvalues. You may inspect any of the three eigenfunctions by selecting **Spectrum | See Wave Functions**.

- **Select the menu item Spectrum | Probability Cloud** and enter the number of the level you want. The program will display the probability density in the x, z plane as a scatter diagram.

- Up till now the program has been working with an orbital angular momentum quantum number of $l = 0$. **Select the menu item Ang Mom | Set Orbital Ang Mom, 1**. Set the new value to 1.

- **Select the menu item Ang Mom | Display Spherical Harmonic**. A three-dimensional drawing of the spherical harmonic corresponding to $l = 1, m + 0$ will appear.

- Find the eigenvalues again, as described above. Examine their shapes and probability distributions as before.

Wave Function Properties

Select the menu item Parts | Part 2: Wave Function Properties. This part of the program can be used to calculate various overlap integrals involving the eigenfunctions you found in part 1, or linear combinations of them.

- **Click on the menu item Psi 1** and select one of the quantum numbers shown. That eigenfunction will appear in the uppermost viewport.

- **Click on the menu item Psi 2** and select a different quantum number. That eigenfunction will appear in the second viewport, and the *product* of the two eigenfunctions you chose will appear in the third viewport.

- **Select the menu item Integrate**. The area under the product function will be calculated. Since you chose two different eigenfunctions they should be orthogonal to one another. Therefore the integral should vanish, to the accuracy of the calculation.

- **Select the menu item Operator | r**. The second viewport will now contain the product of r with the second eigenfunction. The product in the third viewport will also change accordingly. Selecting **Integrate** again will calculate the matrix element of r between the two eigenstates.

A.6   *Walk-Through for* Ident *Program*

Start the program, then choose **File | About Program** for a brief description of terms like joint probability density and separation probability density. (A thorough description is given in section 7.4.)

- **Particles far apart.** When the program starts up, it shows the joint and separation probability densities corresponding to a (nonsymmetrized) bivariate Gaussian wave function. Choose **Symmetry | Compare Symm. and Antisymm.** to see the probability densities associated with the symmetrized and antisymmetrized wave functions arising from this bivariate Gaussian. You will notice that these two look precisely the same! This is because the two particles are far apart.

- **Particles close together.** To examine a situation in which the symmetric and antisymmetric probability densities differ, choose **Wave func | Bivariate Gaussian** and set the mean values of x_1 and x_2 to -0.05 nm and $+0.05$ nm, respectively. This produces a wave function in which the two particles are quite close relative to their associated position uncertainties. Now the two probability densities are quite different. They illustrate the general principle that if the wave function for two identical particles is symmetric, the two particles tend to huddle together; if it is antisymmetric, they tend to spread apart. This is reflected in both the contour plots of joint probability density and in the plots of separation probability density.

- **Other wave functions.** Examine some of the wave functions available through **Wave func | Energy Eigenstate** or **Wave func | Combination of Energy States**. Notice that the huddle together *vs.* spread apart principle applies in all cases.

- **Nonidentical particles colliding.** Reset the program to its initial configuration by choosing **Reset | Reset to Defaults**, then set the particles into motion with **Run**. You are watching two nonidentical particles collide ... but the particles don't interact, so they pass right through each other! You can see this most easily by watching the separation probability density: Initially, it is very likely that the particles are far apart. Then the mound in $PD_{sep}(s)$ moves left toward the origin, crashes into zero (as the particles pass through each other), and finally moves away to the right again as the particles pull away from the collision.

- **Identical fermions colliding.** Go back to the starting point with **Reset | Reset this Run**, but now set **Symmetry | Antisymmetrized**. Upon choosing **Run**, you will see the mound of $PD_{sep}(s)$ moving left, but it stops and begins to move right again *before* it reaches the origin. The two particles never sit right on top of each other. I cannot emphasize enough that these are *noninteracting* particles, and that this apparent repulsion is just that ... apparent. It is not due to any force, it is not due to any term in the Hamiltonian: it is due only to the antisymmetrization requirement of the interchange rule.

- **Identical bosons colliding.** Run the same collision over again with **Symmetry | Symmetrized**. You may be able to see that now there is an effective attraction between the two noninteracting identical particles, although this is more difficult to detect. The best way to see it is by starting over again with **Symmetry | Compare all Three** and with **Plot What | Show Mean Separation**.

A.7 *Walk-Through for* Scattr3D *Program*

The initial screen comes up with the default potential, a spherical barrier.

- **Select Hard Sphere from the Potential menu.** Accept the default parameter values by clicking OK or pressing ⟨**Enter**⟩. The size of the sphere is determined by r_{min}, for which the default value is 1 nm.

- **Select Total Cross Sections from the Compute menu.** In the input screen that appears, enter an angular momentum range of 0 to 0, so as to include only the S-wave, and enter an energy range of 0.01 eV to 0.1 eV, with steps (delta E) of 0.002 eV. The plot that develops shows the total S-wave cross section as a function of energy.

- **Press ⟨F4⟩ to rescale the plot.** Set the horizontal range to run from 0.0 to 0.1 nm, and the vertical range from 10 to 15 nm^2.

- **Extrapolate by hand to zero energy.** The value reached should be ~12.5 nm^2, which is approximately equal to $4\pi(r_{\min})^2$.

- **Select New from the File menu.** This resets most of the programs defaults, including the choice of the potential. Decline the opportunity to avoid overwriting the previous calculation before it is saved.

- **Select Particle Type & Mass, Units from the Parameters menu.** In the dialog box then presented, choose the proton as the particle to be scattered.

- **Select Display & Modify Current Choice from the Potential menu.** Set V_0 to -50 MeV and $r_{\max}$ to 4 fm.

- **Select Total Cross Sections from the Compute menu.** Use 0 to 2 as the range of angular momentum, 0.1 to 3.0 MeV as the range of energy, and 0.02 MeV as the energy step. Most, if not all of the results, will not be visible because the default scale is not tuned to this particular calculation.

- **Press ⟨F4⟩ to rescale the plot.** Use the auto scale option to bring the result into view. Press ⟨F4⟩ again and set the lower limit of the vertical scale to zero.

- **Press ⟨F2⟩ to repeat the calculation.** However, restrict the range in angular momentum to a single value. Obtain additions to the plot for $l = 0$, 1, and 2. Identify the angular momentum and approximate energy (~0.7 MeV) of the resonance.

- **Select Phase Shifts from the Compute menu.** Enter an angular momentum of 0 and an energy range of 0.1 to 3.0 MeV, with an energy step of 0.04 MeV.

- **Press ⟨F2⟩ to repeat the calculation.** Enter an angular momentum of 1. It does go through $\pi/2$ near the resonance. However, there is also a numerical artifact.

- **Press ⟨F5⟩ twice to plot the full magnitude of the phase shift.** The discontinuity that becomes apparent is a failure of the algorithm that adjusts the phase shift to be zero at high energy. In this extreme case the algorithm must be given more cycles with which to work.

- **Select Numerical Parameters, Algorithm from the Parameters menu.** Increase $r_{\text{stop}}/r_{\max}$, $r_1/r_{\max}$, and $r_2/r_{\max}$ to 40, 30, and 34, respectively. Then repeat the phase shift calculation.

- Select **Radial Wave Functions** from the **Compute** menu. Choose an angular momentum of 1 and an energy near that of the resonance.

- **Press ⟨F4⟩ to rescale the plot.** Set the horizontal maximum to 160 fm.

- **Press ⟨F5⟩ to plot the free wave function.** The horizontal white line indicates the distance that the free wave function must be shifted to be in phase with the scattered wave function.

- **Select Exit Program from the File menu.** The program will not allow an immediate exit if calculations have been done but not saved. There is instead an opportunity to save the work to a file. This can be done directly with other entries in the **File** menu. Notice that work saved previously, perhaps in preparation for a lecture demonstration, can be recalled with the **Open...** entry.

A.8   Walk-Through CylSym *Program*

The initial screen comes up with the default initial guess, a Coulombic wave function. The default potential is the quadratic Zeeman potential.

- **Press ⟨F2⟩ to begin iterations toward a solution.** Let it run to completion.

- **Select Viewpoint from the Parameters menu.** This provides a means to change the view of the surface plot. The three Euler angles are varied using the sliders, and the size and distance of the plot are changed with hot keys ⟨**F3**⟩ and ⟨**F4**⟩. Any changes are immediately reflected in the small box in the lower left, and are applied to the wave function plot when ⟨**F2**⟩ is pressed. The default viewpoint can be recovered by pressing ⟨**F5**⟩. When finished adjusting the viewpoint, press ⟨**Enter**⟩.

- **Select Oscillator from the Potential menu.** Change w_1, the angular frequency in the transverse direction, to 2 eV/$\hbar$, so that the potential is not isotropic. An analytic solution does exist, of course, and can be used as a check on the program.

- **Select Gaussian from the Initial Wf menu.** This will be the initial guess. Reasonable values for the parameters are 20 nm^{-2} and 10 nm^{-2}.

- **Press ⟨F3⟩ three times.** This causes the program to take three time steps.

- **Press ⟨F5⟩ to change the plot type.** Select a plot of ⟨*H*⟩ versus time.

- **Press ⟨F2⟩ to continue the time steps.** When the points plotted reach the bottom of the plot, press ⟨**F2**⟩ again to stop. The plot will rescale automatically. Press ⟨**F2**⟩ a third time to finish the calculation. (Stopping and restarting the calculation in this way are not necessary for the completion of the calculation, but only for facilitating observation of changes in ⟨*H*⟩.)

- **Compare the result for $\langle H \rangle$ with the exact eigenvalue of 2.5 eV.** The expectation value need not satisfy the variational principle because it is computed numerically, with errors that can result in underestimates of the integrals involved.

- **Select Exit Program from the File menu.** The program will not allow an immediate exit if calculations have been done but not saved. There is instead an opportunity to save the work to a file. This can be done directly with other entries in the **File** menu. Notice that work saved previously, perhaps in preparation for a lecture demonstration, can be recalled with the **Open...** entry.

Appendix B

The Display of Wave Functions

Any treatment of quantum mechanics raises a deep question in scientific visualization: How should one plot a complex-valued function $\psi(x)$ on a computer screen (or on a piece of paper, for that matter)? One answer is to plot only the square magnitude $|\psi(x)|^2$. This is not adequate because the square magnitude does not convey full information about the state: it says nothing, for example, about the mean momentum. Another possibility is to plot the real and imaginary parts of $\psi(x)$—perhaps using two lines of different color. This solution is again less than optimal, because the real and imaginary parts separately have no physical significance. For example, $\mathrm{Re}\{\psi(x)\}$ often oscillates wildly where $|\psi(x)|$ is quite smooth. Indeed, the division between real and imaginary parts of $\psi(x)$ can be altered completely, without changing the physical state, simply by multiplying $\psi(x)$ with a complex constant of modulus unity. The physically important representation of $\psi(x)$ is not as a real and an imaginary part, but as a magnitude and a phase,

$$\psi(x) = A(x)e^{i\phi(x)},$$

where $A(x)$ (magnitude) and $\phi(x)$ (phase) are both real functions.

Everyone is used to representing a real number like $A(x) = |\psi(x)|$ by a length, but how should one represent a phase? Most graphical representations that come to mind suffer from a discontinuity as the phase passes through 2π radians—a discontinuity that is present in the representation but *not* in the physical entity being represented. However, this defect is absent when phase is represented by *color*. There is a continuum of hue (the "color wheel") that follows the rainbow from red to green to violet and then goes back to red again through magenta. In the "color for phase" representation the magnitude of $\psi(x)$ at point x is represented by the height of a line at that point, while the phase of $\psi(x)$ there is represented by the line's color. The CUPS quantum mechanics programs can represent complex wave functions using any of the mechanisms mentioned here, but the default display style is color for phase.

(Because of limitations in the Turbo Pascal language in which these programs are written, only eight colors are used to represent the ideal color continuum mentioned above. To demonstrate the continuous character of the phase, twenty-four would be far better.)

Index